30-SECOND
GENETICS

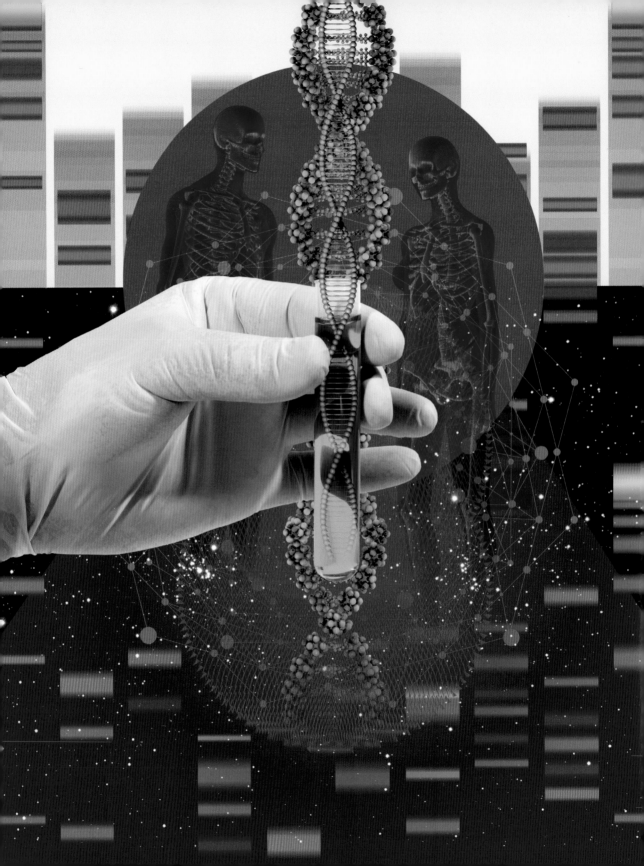

30-SECOND GENETICS

The 50 most revolutionary
discoveries in genetics,
each explained in half a minute

Editors
**Jonathan Weitzman
& Matthew Weitzman**

Foreword
Rodney Rothstein

Contributors
**Thomas Bourgeron
Robert J. Brooker
Virginie Courtier-Orgogozo
Alain Fischer
Edith Heard
Mark F. Sanders
Reiner A. Veitia
Jonathan B. Weitzman
Matthew D. Weitzman**

Illustrations
Steve Rawlings

METRO BOOKS
New York

METRO BOOKS
New York

An Imprint of Sterling Publishing Co., Inc.
1166 Avenue of the Americas
New York, NY 10036

ISBN: 978-1-4351-6613-4

For information about custom editions, special sales, and premium and corporate purchases, please contact Sterling Special Sales at 800-805-5489 or specialsales@ sterlingpublishing.com.

Manufactured in China

10 9 8 7 6 5 4 3 2 1

www.sterlingpublishing.com

Credits:
Publisher Susan Kelly
Creative Director Michael Whitehead
Editorial Director Tom Kitch
Commissioning Editor Kate Shanahan
Project Editors Jamie Pumfrey, Fleur Jones
Designer Ginny Zeal

CONTENTS

FOREWORD
Rodney Rothstein

Genes are the building blocks that assemble in a
myriad combinations to create all forms of life on our planet. Genetics,
and our understanding of genes, touches many aspects of our everyday
lives—including the replication of cells in our bodies and the production
of sperm and eggs for reproduction, as well as complex issues that affect
society such as the creation of genetically modified organisms and the use
of gene therapy technologies. Furthermore, doctors today are beginning
to base diagnoses and treatments on their patients' genetic data.

To some, the word "genetics" conjures up frightening images of
Frankenstein's monster and *Jurassic Park*. To counteract these fears, it is
important for us as scientists to communicate our discoveries and share the
science of genetics in a way that is understandable to the general public. It is
clear that wider knowledge of the underlying principles of genetics will help
demystify this important area of science and also elevate the level of public
discussion of potential ethical issues. My colleagues Matthew and Jonathan
Weitzman, themselves natural genetic clones, together with the authors of
this book present fifty 30-second pieces that distill the essentials of the world
of genetics. These pieces give the reader an insight into how this field grew
from Gregor Mendel's studies of inheritance through to the discovery of
DNA as the genetic material, right up to the present of whole genome
sequencing and into the future of genetic diagnosis and gene therapy.

In the end, genetics should not be feared. As we learn more about
the relationships between genes, as well as their interactions with the
environment, we can look forward to an exciting new era where our
quality of life will be improved. The engineering of healthier food, the
application of synthetic biology to facilitate the production of drugs and
other compounds, and the delivery of better health outcomes through
precision medicine, will all positively affect our lives.

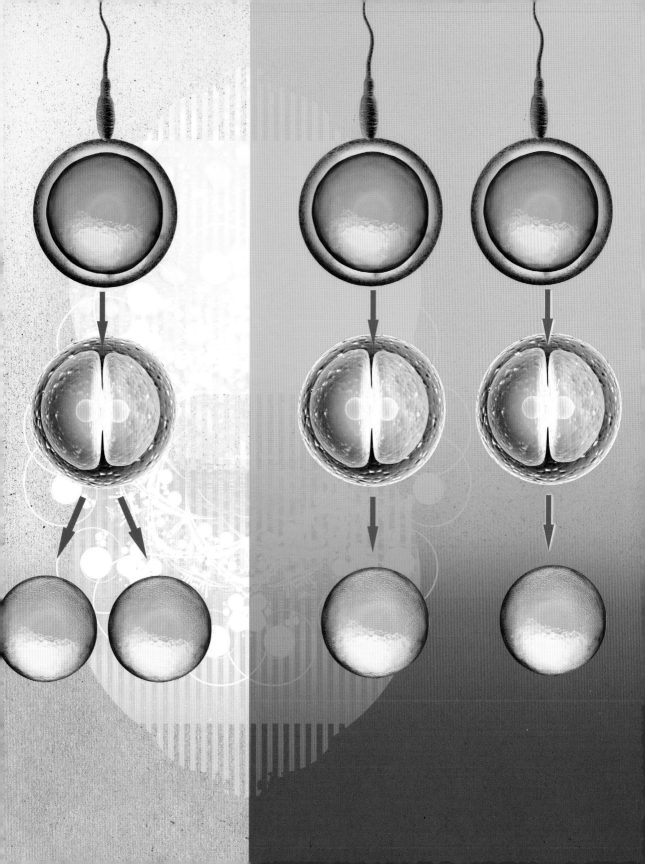

INTRODUCTION
Jonathan Weitzman & Matthew Weitzman

Few scientific disciplines have captured the human imagination like the field of genetics. Perhaps we are fascinated by it because it tackles questions at the very essence of who we are and what gives each of us our personal identities: What explains why we look like our parents? What makes us different from our brothers and neighbors? And what will we pass on to the next generation? These questions are as old as humankind, but a century ago a science was born that would provide unexpected insights, make unprecedented progress, and challenge the ways we think about heredity.

The twentieth century was a whirlwind roller coaster ride, with biomedical promises and ethical challenges at every turn. The painstaking observations of Gregor Mendel allowed him to define some basic rules for the inheritance of characteristics (or "traits"). The rediscovery of his work at the turn of the century set the stage for exploring what is transmitted from one generation to another and how traits are determined. This new science would require new words to describe it. The British biologist William Bateson coined the term "genetics" (from the Greek "to give birth") to describe this new science of heredity in a personal letter in 1905 and then a year later publicly at the Third International Conference on Plant Hybridization in London. Soon the words "gene" and "genome," "genotype" and "phenotype" were born. Strong personalities abounded in this new field, all keen to uncover the mysteries of heredity. There would be moments of great achievement, like the ingenious cracking of the genetic code, or the discovery of the double helix—the beautifully simple structure of the DNA molecule that would become an icon. The twentieth century closed with one of the most exciting challenges in modern biology: a race, comparable to the race to put man on the Moon. Never before had so many geneticists from around the world worked together on a project with such scope and ambition. The international Human Genome Project set out to decrypt the three billion letters that make up the human genome—the Book of Life!

There is no area of biology that has not been profoundly impacted by modern genetics. Much of the progress in the field was due to the extraordinary speed with which new technologies emerged to address challenges. The realization that the way DNA and genes work is the same across the animal and plant kingdoms, opened the door to a menagerie of experimental modeling systems. Discoveries in single-celled bacteria, the common baker's yeast, or the lowly fruit fly provided clues to the underlying rules of genetics. Indeed, the functions of pieces of DNA could be tested by transferring genes from one organism to another. Researchers learned how to sequence, copy, synthesize, and engineer DNA molecules, often exploiting the machines (or "enzymes") that Nature herself had taken millennia to perfect. This extraordinary progress led to breakthroughs in understanding human diseases and the promise of a new type of genetic medicine. But the promises also brought fears and inspired macabre fiction and fantasies.

Progress continues at a breathtaking pace. Human genomes are being sequenced in their thousands, gene therapy is finally correcting errors to save people's lives, and gene editing has reached unprecedented levels of precision. The field of genetics has moved from an esoteric science of abstract concepts, to a series of technologies that will impact our daily lives. In this book we set out to share our excitement at this wonderful adventure and to demystify the science that sometimes hides behind its jargon. The words "gene" and "DNA" have crept into our everyday speech, but it is often unclear what they really mean. It is important to explain what genetics can and cannot say about who we are. The molecules and the enzyme machines that copy, interpret, and protect our genomes are all microscopic, but their impact on society is gigantic and in *30-Second Genetics* we want to equip general readers to participate in the debate about how genetics and genetic information will be used by society and by generations to come.

About this book

In *30-Second Genetics*, experts from around the globe guide us through the jargon of modern genetics from "gene" to "genome," from the deciphering of the genetic code to the sequencing of the human genome. Here specialists demystify the terms and the concepts, and make us wonder at how much we have learned about genetics and how much we still have to discover. *30-Second Genetics* presents each topic in a clear and concise single page. The main paragraph, the **30-Second Theory**, is complemented by the **3-Second Thrash**, which gives a quicker overview—the key facts in a single sentence. And the **3-Minute Thought** fleshes this out, adding intriguing aspects of the subject. Each chapter also contains the biography of a pioneer in the field—the men and women who contributed to our understanding of modern genetics. The book begins with a presentation of the historical and conceptual foundations of this new science. It then plunges into the details, first by explaining the roles of chromosomes and cells through to the level of genes and genomes, before discussing the emerging field of epigenetics, which studies genetic effects that are not encoded in the DNA sequence of an organism. The Health & Disease chapter places these molecular events in the context of physiology and the bodily processes that are associated with disease. No discussion of genetics would be complete without a description of the progress in technologies and experimental approaches. The book ends with some predictions of how these technologies might impact our lives, and the therapeutic promises these approaches may provide in the near future.

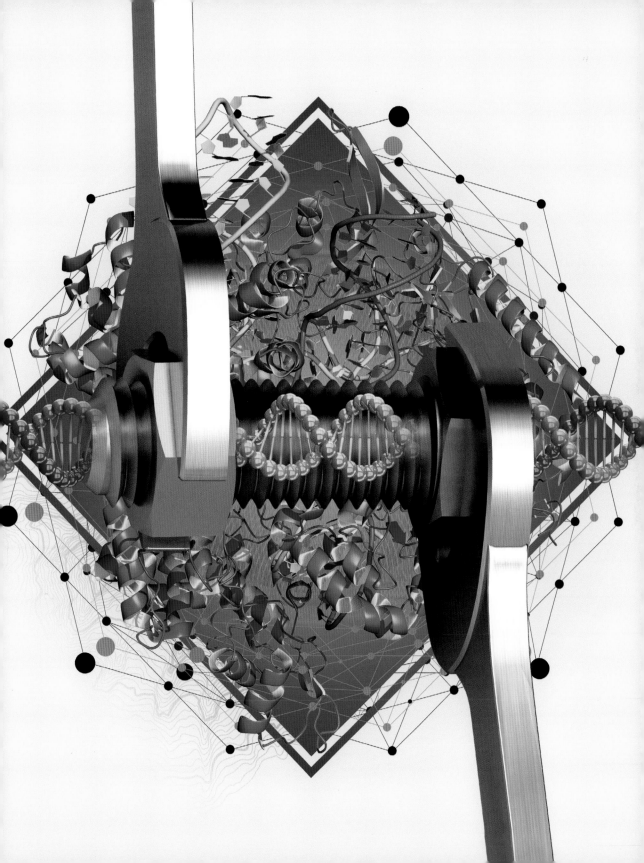

HISTORY & CONCEPTS

alleles Alternative variant forms of a gene that result from a mutational change in DNA sequence or expression of the gene. Alleles can be recessive, meaning they only have an effect when there are two copies, or dominant, where a single copy is enough to have an effect.

amino acids Water-soluble organic compounds that are the building blocks to make proteins. There are around 20 amino acids. Ten cannot be made by the human body and so must be included in the diet: they are called "essential amino acids."

chromosomes Long strings of double-stranded DNA that carry genes and the genetic information. In eukaryotic cells (those with a discrete nucleus) the chromosomes are in the nucleus and composed of DNA, some RNA, and proteins. A prokaryotic cell (one without a discrete nucleus) has a single chromosome made of DNA and a very small amount of protein.

codon Genetic information is coded in triplets of DNA that encodes triplets of messenger RNA (mRNA) nucleotides. Each string of three mRNA nucleotides is called a codon and each codon codes for a different amino acid used to make the corresponding protein.

double helix The double-stranded structure of DNA. The two strands of DNA are wrapped around each other like a twisted cord.

DNA Deoxyribonucleic acid is a long molecule that carries the genetic information and transmits inherited traits. DNA is found in the cells of all prokaryotes and eukaryotes.

gametes Specialized cells for sexual reproduction. Male gametes are sperm cells and female gametes are egg cells.

genes Units of heredity, located on a chromosome. Genes consist of DNA, except in some viruses where they are made of RNA. Particular genes control specific cellular processes—for example, genes can control cell division and cell suicide.

genome Complete set of genetic material in an organism or a cell. Genomics is the study of an organism's genome, focusing on its evolution, function, and structure.

locus (pl. loci) Position of a gene on a chromosome. Different alleles of the same gene occupy the same locus.

mutation Change in DNA sequence or gene structure that can result from substitutions of bases of DNA or the rearrangement, deletion, or addition of sections of genes or chromosomes.

nucleotides Building blocks used to make DNA or RNA. Strings of nucleotides are called nucleic acids. In DNA there are four nucleotides (referred to by the letters T, C, G, and A) and in RNA there are four ribonucleotides (U, C, G, and A). Nucleotides are also called "bases." DNA bases can be paired: A pairs with T, and C pairs with G.

polymer Long molecule made up of simpler building blocks (monomers). DNA is a polymer chain made up of strings of nucleotides. Proteins are polymer chains made up of strings of amino acids. Proteins are sometimes called polypeptide chains.

replication Process of exactly copying DNA, usually for doubling the DNA before a cell divides to make two new cells. Replication involves an enzyme machine called DNA polymerase that copies each strand of the DNA to make a precise complementary copy of the DNA molecule.

RNA Ribonucleic acid, a molecule that is made in all living cells and is important for synthesizing proteins and for regulating genes. RNA is normally made by copying one strand of the DNA. Messenger RNA (mRNA) is a copy of DNA that contains the information to make a protein. In some viruses, RNA rather than DNA functions as the carrier of genetic information.

species Group of organisms whose members can interbreed and produce fertile offspring. "Species" is the eighth category in the scientific classification system, beneath "genus."

transcription Process for turning DNA genetic information into RNA. This is done by an enzyme machine called RNA polymerase, which builds an RNA polymer using the DNA as a template.

translation Process for making proteins using genetic information in messenger RNA. The ribosome is a large protein machine that moves along the mRNA and reads mRNA codons. The ribosome links amino acids to make a protein chain.

MENDEL'S LAWS OF HEREDITY

the 30-second theory

Gregor Mendel uncovered the laws of heredity while experimenting with garden peas. He bred plant lineages in isolation for many generations so that their offspring had various identical visible properties. Then he crossbred plants with different visible properties, for example plants with purple flowers with plants with white flowers. In the first generation, only plants with purple flowers were obtained. After crossbreeding these plants again he observed that one-quarter of the new plants had white flowers and three-quarters of them had purple flowers. To explain this, Mendel concluded that they resulted from the transmission of pairs of factors, which determined visible traits according to the laws of probability. The character *purple flowers*, which dominates in the first generation, is said to be dominant (P) compared to *white flowers*, which is recessive (p). In humans, blue eye color is recessive and brown eye color is dominant. Mendel's factors are now known as "alleles," which are variations in the DNA sequence that specifies traits. By extension, one can speak of dominant or recessive alleles. These alleles are alternative sequences of a *locus* (Latin for "place"), which in many cases can be loosely equated to a gene. More than two alternative alleles can exist in a population.

Two recessive alleles (pp) result in a recessive trait, whereas two dominant alleles (PP) or a dominant and a recessive allele (Pp or pP) will result in the organism expressing a dominant trait.

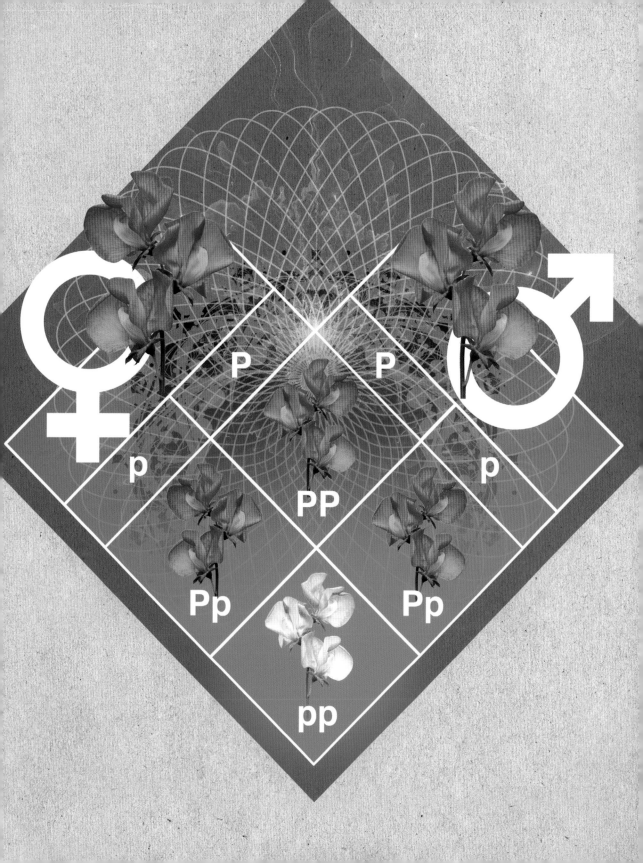

DARWIN & THE ORIGIN OF SPECIES

the 30-second theory

Where do we come from? Why do we have limbs and eyes? Such questions were considered to be outside the realm of science until Charles Darwin published his magnum opus *On the Origin of Species* in 1859. Darwin's view of life is now called the theory of evolution. Briefly, it states that some of the traits that differ between individuals within a population can be transmitted to the next generation. The individuals who are best adapted to the environment in which they live are most likely to survive, reproduce, and pass on their heritable characters to future generations. In this way, populations change over time, adapting to their environment; this can eventually lead to new species. Darwin's ideas conflict with the intuition that humans are distinct from other animals or with the belief that species remain invariant over time. His book ignited huge philosophical and religious debates, some of which are still ongoing. The discovery of genes, genetics, and DNA in the 1920s–60s provided new support for Darwin's theory. This led to today's modern theory of evolution, which is central to our understanding of the living world.

RELATED TOPIC
See also
GENES & ENVIRONMENT
page 78

3-SECOND BIOGRAPHIES
ALFRED RUSSEL WALLACE
1823–1913
British naturalist who conceived the theory of evolution at the same time as Darwin

THEODOSIUS DOBZHANSKY
1900–75
Russian-American geneticist, famous for stating that "nothing in biology makes sense except in the light of evolution"

JERRY COYNE
1949–
American biologist who actively promotes the theory of evolution in books and blogs

30-SECOND TEXT
Virginie Courtier-Orgogozo

Darwin's theory of evolution by natural selection is one of the most revolutionary ideas in the history of science.

3-SECOND THRASH
Darwin's book is a masterpiece of observation and creative thinking; it profoundly changed how humanity conceives its origins.

3-MINUTE THOUGHT
Like other scientific explanations, the theory of evolution is challenged and refined by new facts. Although the core of the theory presented by Darwin is still valid today, certain parts have been refuted (for example, the diversification of species resembles a meshwork rather than a branching tree, as he suggested) and others (such as the origin of life) remain enigmatic.

COLUMBA LIVIA or ROCK-PIGEON

II.	GROUP III.						GROUP IV.
4.	5.	6.	7.	8.	9. SUB-GROUPS.	10.	11.

Persian
Tumbler

Lotan
Tumbler

Common
Tumbler

Tronk.

Java
Fantail.

Turbit.

Barb.   Fantail.   African   Short   Indian   Jacobin.
                                Frill-
                     Owl. faced   back.
                          Tumbler.

English Pouter.
Laugher.
Trumpeter.

Dove-cot pigeon.
Swallow.
Spot.
Nun.

DNA CARRIES THE GENETIC INFORMATION

the 30-second theory

The history of the discovery of deoxyribonucleic acid (DNA) can be traced back to the work of Friedrich Miescher, who isolated a substance he termed "nuclein" from the nuclei of white blood cells in the late 1880s. This substance is composed of proteins and what we now know as DNA. Its former name, coined by Richard Altmann, was the generic "nucleic acid." Later on, Frederick Griffith showed that a substance derived from disease-causing (pathogenic) bacteria could change non-pathogenic bacteria into virulent forms. Griffith's experiment was continued by Oswald Avery, Colin MacLeod, and Maclyn McCarty. They destroyed everything but the DNA of pneumonia-producing bacteria. After this drastic treatment, DNA could still transform non-pathogenic into pathogenic bacteria. Only the destruction of DNA prevented this transformation and this demonstrated that it was the DNA that carried the genetic information. In the meantime, Phoebus Levene had identified the components of DNA: the bases adenine, cytosine, guanine, thymine, a sugar molecule, and a phosphate group. All these discoveries paved the way for the unraveling of the chemical structure of DNA by Rosalind Franklin, Maurice Wilkins, James Watson, and Francis Crick in the early 1950s.

3-SECOND THRASH
Experiments in the 1940s formally demonstrated that DNA is the molecule that carries the genetic information of most known organisms.

3-MINUTE THOUGHT
The history of the discovery of DNA and its structure is tainted by injustice. The results of Avery, MacLeod, and McCarty were largely unrecognized and unaccepted. In another famous example, Watson and Crick built their famous double-helix model based on pictures of DNA's structure that were obtained by Rosalind Franklin and Maurice Wilkins. Franklin died at 37 and her vital contribution to the story has been downplayed until recently.

RELATED TOPICS
See also
THE DOUBLE HELIX
page 22

CRACKING THE GENETIC CODE
page 24

THE CELL NUCLEUS
page 36

3-SECOND BIOGRAPHIES
JOHANNES FRIEDRICH MIESCHER
1844–95
Swiss physician and biologist who first identified nuclein and nucleic acids

OSWALD AVERY, JR.
1877–1955
Canadian-born American physician who demonstrated that DNA is genetic material

PHOEBUS LEVENE
1863–1940
American biochemist who identified the components of DNA

30-SECOND TEXT
Reiner Veitia

DNA's basic components include the four bases: adenine, cytosine, guanine, and thymine.

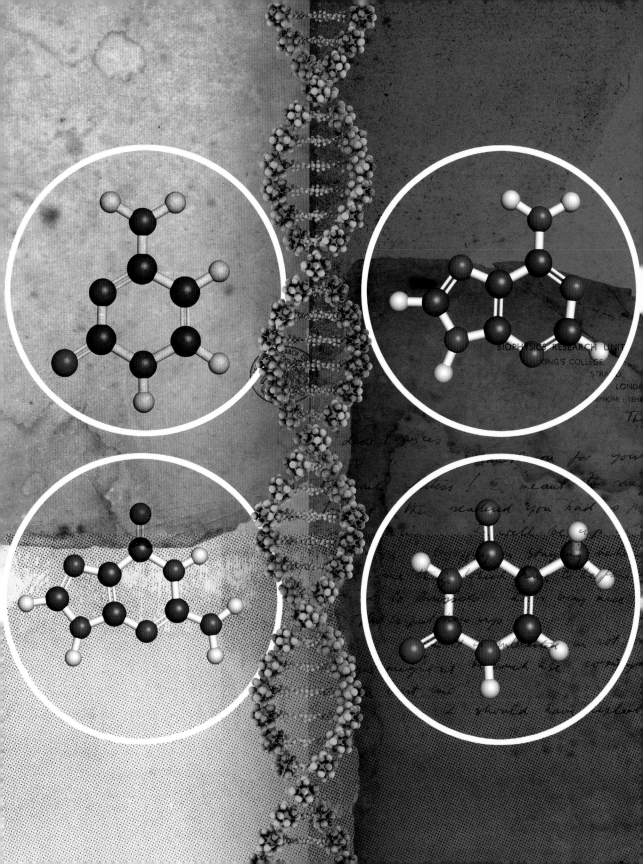

THE DOUBLE HELIX

the 30-second theory

The function of DNA depends
on its structure. DNA is composed of building
blocks called nucleotides, which consist of a
deoxyribose sugar, a phosphate group, and one
of four bases: adenine (A), thymine (T), guanine
(G), and cytosine (C). Nucleotides link together
in long chains known as polymers, and each
is specifically paired with a nucleotide on the
opposite strand: A always bonds with T, and
C bonds with G. In the early 1950s there was
a race to work out how these base pairs fit
together into a three-dimensional structure.
Rosalind Franklin, working with Maurice Wilkins
at King's College London, beamed X-rays
through crystals of the DNA molecule to gain
insights into its structure. This X-ray diffraction
technique produced an image that suggested
DNA molecules form a helical shape. James
Watson and Francis Crick, who were working
at the Cavendish Laboratory in Cambridge, saw
this image and realized that it provided a critical
clue to the structure of DNA. They developed
a chemical model for the DNA molecule, and in
1953 were the first to propose that its structure
is that of a double helix. Further research into
the structure revealed the mechanism for base
pairing, and explained how genetic information
can be stored and copied in living cells.

RELATED TOPICS
See also
DNA CARRIES THE
GENETIC INFORMATION
page 20

THE CENTRAL DOGMA
page 28

WHAT IS A GENE?
page 56

3-SECOND BIOGRAPHIES
FRANCIS CRICK
1916–2004
British biophysicist who
co-discovered DNA's structure
with James Watson

ROSALIND FRANKLIN
1920–58
English chemist who generated
the crucial X-ray diffraction
images of the DNA molecule

JAMES WATSON
1928–
American biologist and
co-discoverer of DNA's structure

30-SECOND TEXT
Matthew Weitzman

3-SECOND THRASH
The discovery of the
molecular structure of
DNA was a landmark
moment in genetic and
molecular biology research.

3-MINUTE THOUGHT
Francis Crick and James
Watson first described
the structure of a
double-stranded DNA
molecule as a "double helix"
in the journal *Nature* in
1953. Two linear strands
run in opposite directions
and are connected into a
twisted helical structure.
The sequence of the bases
in each strand makes a
digital code that carries
the instructions for life.

*Watson and Crick won
a Nobel Prize in 1962
for discovering DNA's
double helix structure.*

CRACKING THE GENETIC CODE

the 30-second theory

Working out the encryption rules to crack a code is the romantic job of spies and secret agents. But researchers also had to be pretty cunning to work out how information encoded in the DNA sequence is turned into the string of amino acids in proteins. The genetic code is the rulebook for translating DNA information into protein information. The code is extremely similar among all living organisms. As there are four "letters" called nucleotides in DNA (A, G, C, and T) and these need to code for 20 amino acids, it quickly became clear that the code had to involve at least three letters. That gave 64 possible combinations, but which one corresponded to which amino acid? In the 1960s researchers performed pioneering experiments to prove that a triplet code (called a codon) was responsible for each amino acid. The code-cracking break came when they used a cell-free system and put in RNA molecules with a long string of just one letter. Using a synthetic chain of nucleotides called "poly-uracil" they deduced that UUU was the code for the amino acid phenylalanine. Then it was just a matter of working out the other combinations. Today, with the full table of 64 combinations, any student can predict a protein sequence from DNA code.

3-SECOND THRASH
DNA information in genes is organized into triplet codons, where three letters of the DNA encode for one amino acid in the resulting protein.

3-MINUTE THOUGHT
Because there are 64 possible combinations of three letters, there is an inherent redundancy. For example, there are four codons that correspond to the amino acid Alanine (GCU, GCG, GCA, or GCG). This means that the third letter can change without affecting the code. This is called a "silent mutation." Researchers use the word "degeneracy" to refer to this redundancy in the genetic code.

RELATED TOPICS
See also
DNA CARRIES THE GENETIC INFORMATION
page 20

THE CENTRAL DOGMA
page 28

DNA SEQUENCING
page 126

3-SECOND BIOGRAPHIES
GEORGE GAMOW
1904–68
Ukrainian-American physicist who hypothesized that the genetic code might be made up of three letters (nucleotides) in DNA

MARSHALL W. NIRENBERG
1927–2010
American biochemist who cracked the first codon to begin solving the puzzle of the genetic code

30-SECOND TEXT
Jonathan Weitzman

The three-letter DNA code contains the information for making proteins.

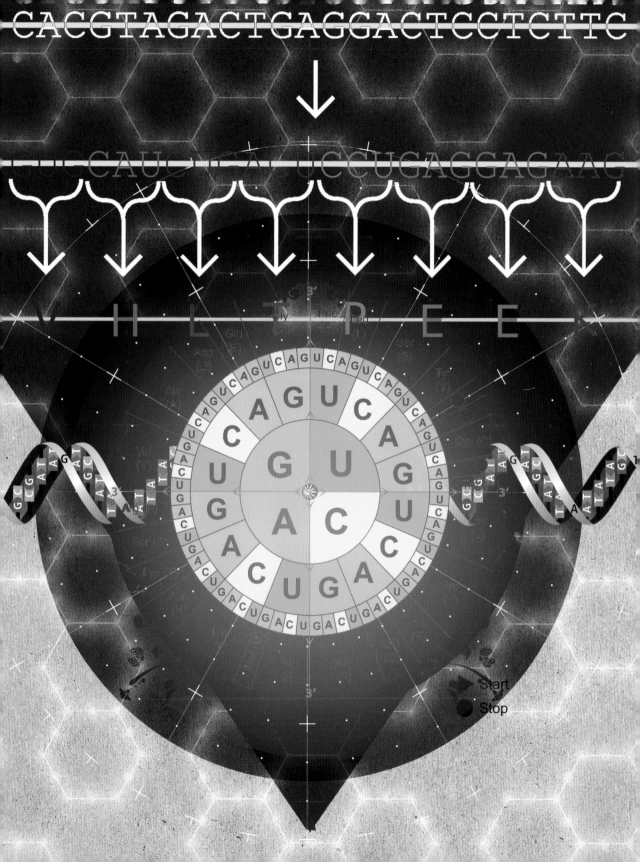

July 25, 1920
Born in London, England

1938
Studies at Newnham College, University of Cambridge, England

1946
Earns a PhD in Physical Chemistry at the University of Cambridge

1946–50
Learns X-ray crystallography in crystallographer Jacques Mering's laboratory in Paris

1951–53
While working in John Randall's laboratory at King's College, London, produces X-ray photographs of DNA

1952
Her student Raymond Gosling takes the landmark "Photograph 51" of DNA

1953
The April issue of the journal *Nature* publishes three papers on the structure of DNA; one from Franklin's team, one from Wilkins' team, and the third by Watson and Crick

1954–56
Works on tobacco mosaic virus and the poliovirus

April 16, 1958
Dies of ovarian cancer

1962
Crick, Watson, and Wilkins share the Nobel Prize in Physiology or Medicine

2003
The Royal Society establishes the Rosalind Franklin Award for outstanding contributions to science, engineering, or technology

ROSALIND FRANKLIN

Rosalind Franklin was born in

1920 in Notting Hill, London, into an affluent Jewish family. She excelled at sciences and attended St. Paul's Girls' School, one of the few girls' schools in London that taught chemistry and physics. She decided to become a scientist when she was 15, against her father's wishes. Enrolling at Newnham College, Cambridge, England, she graduated in 1941 with a degree in chemistry, and later received a PhD in physical chemistry from the same university.

Franklin played a key role in what is perhaps the greatest achievement of molecular biology: the discovery of the structure of DNA. Her story is one of competition, expertise, and controversy, told one way in James Watson's book *The Double Helix* and in a very different way in Anne Sayre's book, *Rosalind Franklin and DNA,* or Brenda Maddox's biography *Rosalind Franklin: The Dark Lady of DNA.*

In the autumn of 1946, Franklin was appointed at the Laboratoire Central des Services Chimiques de l'Etat in Paris, where she learned the method of X-ray diffraction from crystallographer Jacques Mering. With this technique she created pictures of DNA molecules using X-rays when she returned to England in 1951 as a research associate in John Randall's laboratory at King's College, London.

Maurice Wilkins also worked in the same laboratory. He showed molecular biologist James Watson one of Franklin's crystallographic pictures of DNA. Having seen her data without her knowledge, Watson and his colleague Francis Crick used it to solve the structure of DNA. As Watson candidly wrote, "Rosy, of course, did not directly give us her data. For that matter, no one at King's realized they were in our hands." Watson and Crick used Franklin's photo when they published their findings in *Nature* in 1953. Franklin's photograph has been called one of "the most beautiful X-ray photographs of any substance ever taken."

After leaving King's College, Franklin focused her efforts on the study of viruses, including tobacco mosaic virus and poliovirus. In the summer of 1956, Franklin became ill with ovarian cancer. She died less than two years later in Chelsea, London, at the age of just 37.

Four years after her death, James Watson, Francis Crick, and Maurice Wilkins were awarded the Nobel Prize in Physiology or Medicine. As the prize is not awarded posthumously, Franklin did not receive the credit she deserved for her contribution. The recent increase in awards and buildings that carry her name has reinstated Franklin as one of the forgotten women heroes of genetics.

Robert Brooker

THE CENTRAL DOGMA

the 30-second theory

3-SECOND THRASH
The central dogma of molecular biology describes the flow of genetic information: DNA has the information to make RNA, which has the information to make protein.

3-MINUTE THOUGHT
The central dogma describes the faithful transfer of genetic information from nucleic acid to protein. DNA information can be copied into messenger RNA through the process of transcription. Proteins are synthesized using the information in mRNA as a template in the process termed "translation." Today we know that there is a good deal of information in DNA and RNA that is not used to make proteins but instead regulates genome function.

The "central dogma" of molecular biology describes the transfer of information from DNA to RNA to protein. It was first articulated by Francis Crick to explain information transfer from one polymer molecule with a defined "alphabet" to another. In this way the ordered sequence information is faithfully maintained. Replication is the process of copying DNA, in which information from the parent DNA strand is transferred to the daughter strand. Transcription is the process by which information from DNA is transferred into messenger RNA (mRNA). Translation is the process whereby the sequence of mRNA is used as a template to synthesize proteins. These newly synthesized polypeptide chains are then processed, folded, and modified to form functional proteins. The central dogma states that there is a directional flow of sequence information, and that information cannot be transferred from protein to protein or from protein back to nucleic acid. Discoveries from viruses have shown that RNA can be replicated into RNA, and that RNA can be reverse-transcribed back to DNA, but there is no evidence for reversible transfer of information from proteins to DNA.

RELATED TOPICS
See also
DNA CARRIES THE
GENETIC INFORMATION
page 20

CRACKING THE GENETIC CODE
page 24

NON-CODING RNA
page 90

3-SECOND BIOGRAPHIES
FRANCIS CRICK
1916–2004
British biophysicist and co-discoverer, with James Watson, of the structure of DNA, who coined the term "central dogma" to summarize the flow of genetic information from DNA to RNA to protein

HOWARD TEMIN
1934–94
American virologist who discovered reverse transcriptase, the enzyme that turns viral RNA into proviral DNA

30-SECOND TEXT
Matthew Weitzman

Genetic information is transcribed from DNA to RNA, and then from RNA to protein.

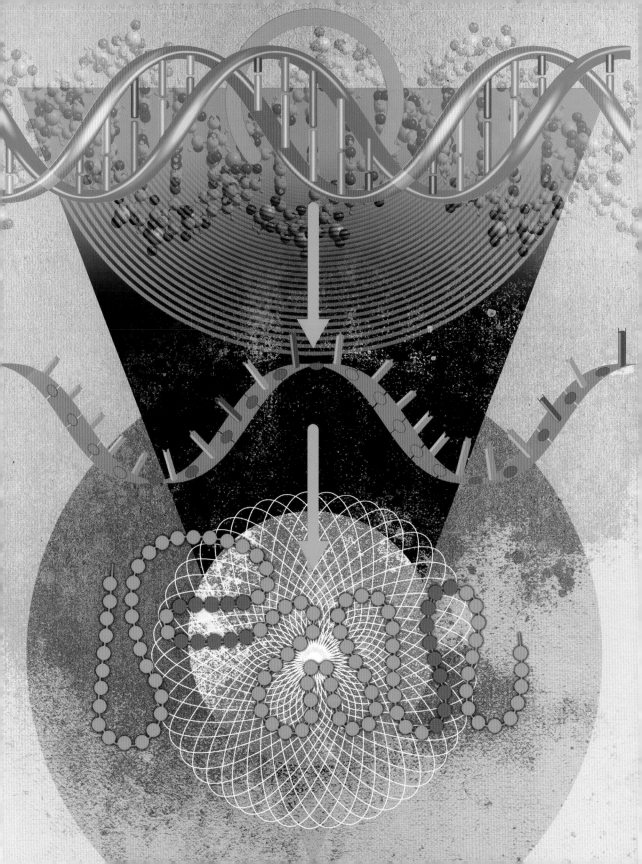

THE HUMAN GENOME PROJECT

the 30-second theory

3-SECOND THRASH
The Human Genome Project sequenced all the DNA letters in humans and made the sequence freely available for all to study.

3-MINUTE THOUGHT
The generation of the first human genome took 13 years, involved thousands of researchers around the world, and cost billions of dollars. DNA sequencing technology continues to increase in speed and accuracy and decrease in cost and time required. Today it takes just hours to sequence a human genome and costs less than $1,000.

The Human Genome Project

is probably the biggest collaborative project ever undertaken by biologists. It is biology's equivalent of the Apollo program that took humans to the moon. The genome is all the DNA that contains all the genes in a cell. Laboratories from around the world joined forces to map and understand all the genes in humans. Following much debate in the 1980s, the US National Institute of Health (NIH) launched the Human Genome Project in 1990, expecting it to last at least 15 years. The project began by making a map of the 23 human chromosomes. This was followed by ordered sequencing of human DNA in research centers around the world. In 1996, the leaders of the project drafted the "Bermuda Principles" to encourage the sharing of all genetic information. The rapid increase in the efficiency and speed of DNA sequencing technology accelerated the project. Sequencing the human genome became a race in 1998, when the private company Celera Genomics set out to sequence the genome at the same time. The first draft of the human genome sequence was published by both the public and the private projects in 2001. The complete sequence published in 2003 showed there are around 20,000 human genes encoded in a genome that contains three billion letters.

RELATED TOPICS
See also
WHAT IS A GENE?
page 56

GENETIC MAPS
page 124

DNA SEQUENCING
page 126

3-SECOND BIOGRAPHIES
JAMES WATSON
1928–
American molecular biologist and co-discoverer of the DNA helix, who was the first person to have his genome sequenced

J. CRAIG VENTER
1946–
American biotechnologist who founded Celera Genomics, the company that raced the Human Genome Project to the finish line

FRANCIS COLLINS
1950–
American geneticist who led the Human Genome Project at the NIH

30-SECOND TEXT
Jonathan Weitzman

The Human Genome Project is one of the largest biology projects of all time.

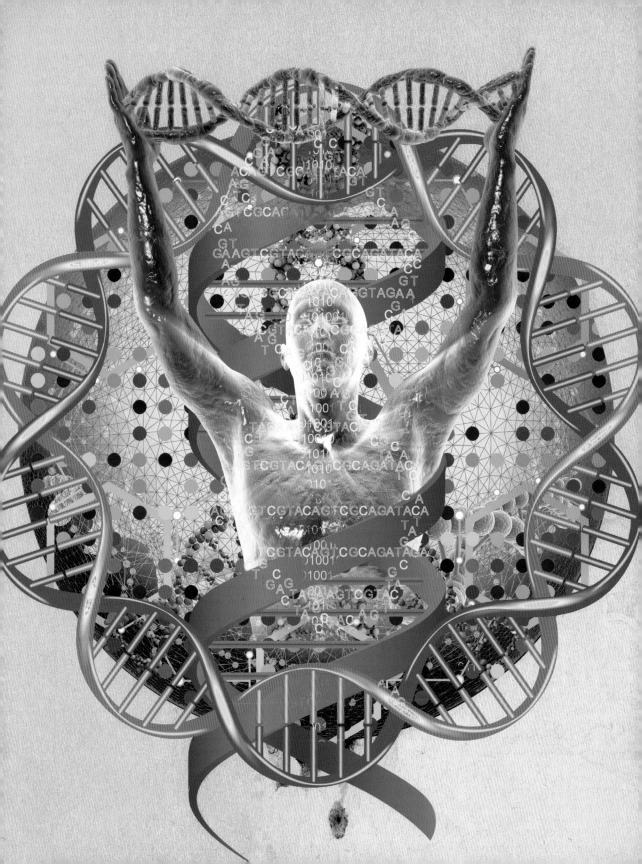

CHROMOSOMES & CELLS

CHROMOSOMES & CELLS
GLOSSARY

alleles Alternative variant forms of a gene that result from a mutational change in DNA sequence or expression of the gene. Alleles can be recessive, meaning they only have an effect when there are two copies, or dominant, where a single copy is enough to have an effect.

ATP Adenosine 5′-triphosphate, a molecule composed of the nucleoside adenine, the sugar ribose, and three phosphate groups. ATP is a small compound that is the main carrier and store of energy in the cell.

centromere Condensed region of the chromosome that holds together the chromatids when cells divide by mitosis. The centromere is the site of assembly of the kinetochore so that the chromosomes can be separated into two daughter cells.

checkpoint proteins Group of proteins that monitor and control the progression of the eukaryote cell through the cell cycle. At several critical points in the cell cycle, the checkpoint proteins ensure that the conditions are met to move on to the next phase of the cycle. They behave like a precise quality-control assessment to allow cells to divide.

chromatid Copy of a chromosome formed when DNA is copied for cell division. The pair of two chromatids (called "sister chromatids") are held together by the centromere.

chromatin Complex formed along the DNA in eukaryote cells. Chromatin is composed of proteins called histones as well as non-histone proteins. The structure of the chromatin plays a key role in regulating gene expression.

cytokinesis Division of the cytoplasm of the parental cell to generate two daughter cells. Distinct from the division of the nucleus (mitosis or meiosis).

cytoplasm Total contents of a cell contained within the outer cell membrane. In eukaryote cells the cytoplasm is all the contents outside of the nucleus.

eukaryote Organism composed of one or many cells with a distinct nucleus and cytoplasm. There are also living cells with no nucleus, such as bacteria, called "prokaryotes."

histones Family of small proteins that are associated with DNA in eukaryotic cells. Many are grouped together in balls of proteins called nucleosomes. This packaging of DNA by histones helps to organize the genome and control gene expression.

kinetochore Complex of proteins that assembles on the centromere during mitosis. The kinetochore creates a link between the chromosomes and the microtubules, so that the chromosomes can be dragged to opposite poles of the dividing cell.

mitochondrion (pl. mitochondria) Organelles in the cytoplasm of eukaryotic cells that produce most of the chemical energy of the cell in the form of ATP, earning them the title "the powerhouse of the cell." They are bound by a double membrane and contain their own genome, referred to as mitochondrial DNA (mtDNA). Dysfunction in mitochondrial function or mutations in the mtDNA can lead to major human diseases involving the metabolism and are collectively called mitochondrial diseases.

mitosis and meiosis Specific types of cell division in eukaryotic cells. Mitosis involves the condensing of the DNA into visible chromosomes and division of the nucleus to create two identical daughter cells with the same amount of DNA and chromosomes as the parent cell. Meiosis involves two rounds of nuclear division to produce four cells, each with half the amount of DNA. Meiosis generates eggs and sperm.

nuclear matrix Network of fibers inside the cell nucleus that organize genetic information within the cell.

nuclear pores Protein complexes that span the envelope surrounding the nucleus. There may be as many as 2,000 in the nucleus of vertebrate cells. They allow the transport of molecules in and out of the nucleus.

organelle Specialized substructure inside the cell with a specific function. Examples of organelles within eukaryotic cells include the mitochondria that generate energy for the cell and chloroplasts that perform photosynthesis in plant cells.

shelterin Protein complex that protects the telomeres at the end of chromosomes from DNA repair mechanisms. In the absence of shelterin, the unprotected telomere can look like broken DNA to the cell, leading to catastrophic attempts to repair it.

telomere Special structure at the end of the chromosome. In eukaryotic the enzyme telomerase is needed to maintain telomeres at each cell division.

THE CELL NUCLEUS

the 30-second theory

The nucleus is like the brain

(or "headquarters") of the eukaryotic cell: it stores information, receives external and environmental messages, and controls the appropriate responses. The cell nucleus is a compartment inside a cell that contains the chromosomes surrounded by a double-membrane called the "nuclear envelope." Nuclear pores provide a passageway for the movement of chemicals in and out of the nucleus. In most human cells the nucleus contains 46 chromosomes, made up of DNA and proteins that compact the chromosomes so they can fit inside the nucleus. The primary function of the cell nucleus is the protection, organization, replication, and expression of the genetic material. The cell nucleus also processes information it receives from other parts of the cell and makes decisions that will alter the cell's structure and function. For example, if there are many nutrients around the cell it will send signals to the cell nucleus and the nucleus will respond by turning on genes that are necessary for digesting nutrients. The nucleus is a control center that stores all of the information necessary for correct cell structure and function.

RELATED TOPICS

See also
CHROMOSOMES
& KAROTYPES
page 38

CELL DIVISION
page 50

GENOME ARCHITECTURE
page 72

3-SECOND BIOGRAPHIES

ANTONIE VAN LEEUWENHOEK
1632–1723
Dutch microbiologist who developed improvements in microscopy that allowed him to observe the nucleus in the blood cells of salmon

FELICE FONTANA
1730–1805
Italian physiologist who is credited with being the first to observe a subnuclear structure called the nucleolus in the slime from an eel's skin in 1781

30-SECOND TEXT
Robert Brooker

3-SECOND THRASH
The nucleus is the central compartment inside a cell that contains the genetic material.

3-MINUTE THOUGHT
At one time, scientists thought that the chromosomes were randomly distributed within the nucleus, tangled up like a plate of spaghetti. But we now know that protein filaments form a nuclear matrix that organizes the chromosomes within the nucleus. Each chromosome is located in a distinct, non-overlapping chromosome territory, which is visible when cells are colored with dyes that label specific chromosomes.

The nucleus stores genetic information and controls the functions of the cell.

CHROMOSOMES & KARYOTYPES

the 30-second theory

Chromosomes are microscopic particles that bear a cell's genetic material. In bacteria, chromosomes are usually composed of a small circle of DNA and proteins. In more complex organisms with several chromosomes, such as ourselves, the DNA is condensed with proteins called histones. DNA, histones, and other proteins constitute the chromatin. In eukaryotic cells, such as ours, the chromosomes are located inside the nucleus. Humans have 23 pairs of chromosomes, with more than 6 ft 6 in (2 meters) of DNA packed within each cell. Of these pairs, 22 are called autosomes or "homologs." They are similar copies of each other that may serve as a back-up of genetic information. The remaining chromosome pair comprises the X and Y chromosomes, called "sex chromosomes," because they determine the sex of an individual. The entire set of chromosomes is called the karyotype, which is studied during cell division at the moment when DNA has been duplicated. At this instant, each chromosome has two condensed copies of its genetic material known as the chromatids. During cell division, the mother cell transmits one chromatid of each chromosome to the two daughter cells. A failure in this process leads to an abnormal number of chromosomes, as is often seen in cancer cells.

Chromosomes are located inside the cell nucleus. They are formed of condensed strings of DNA and proteins.

MITOCHONDRIA

the 30-second theory

3-SECOND THRASH
Mitochondria have their own chromosome and genes. These produce proteins that work with proteins created by the genes in the cell's nucleus in the synthesis of the cellular energy-producing compound ATP.

3-MINUTE THOUGHT
Mutations of mitochondrial genes can impair ATP production and these mutations cause several hereditary disorders. Mitochondrial diseases are classified as "non-Mendelian" for two reasons. First, these diseases are exclusively matrilineal—they only pass from mothers to children. Second, unlike Mendelian genetic diseases, which involve mutation of one or both copies of a gene, mitochondrial diseases usually appear only after many mutant copies of a gene are present in cells.

Inside the nucleus of human cells
are 46 chromosomes carrying over 22,000 genes, the so-called "nuclear genes." But these are not the only genes in human cells. Outside the nucleus, the cells of animals and plants contain several dozen organelles known as mitochondria, each with multiple copies of their own separate genome. Human cells contain about 100 mitochondria, each containing about five copies of the chromosome—so in total there are 500 copies of each mitochondrial gene. Mitochondrial DNA (mtDNA) encodes different numbers of genes in different organisms. There are 37 genes on the human mitochondrial chromosome, 14 of which encode proteins. Human mitochondrial proteins work with proteins from nuclear genes to produce the essential energy-generating compound ATP (adenosine triphosphate). Sperm do not contribute mitochondria during fertilization—the transmission of mtDNA is exclusively via the mother. Millions of years ago the ancestors of mitochondria were independent single-celled organisms. At that time they probably invaded the ancestral cells of plant and animal cells. Once internalized, the ancient invaders and their host cells slowly evolved a symbiotic relationship that has led these former invaders to become integral to the survival of animal and plant cells.

RELATED TOPICS
See also
THE CELL NUCLEUS
page 36

CELL DIVISION
page 50

DOMINANT & RECESSIVE
GENETIC DISEASES
page 104

3-SECOND BIOGRAPHIES
RICHARD ALTMANN
1852–1900
German pathologist who first identified mitochondria and proposed that they had cellular functions

EUGENE KENNEDY
& ALBERT LEHNINGER
1919–2011 & 1917–86
American biochemists who co-discovered the process that produces the compound ATP in mitochondria

30-SECOND TEXT
Mark Sanders

Mitochondria sit outside the cell nucleus but also contain their own genome.

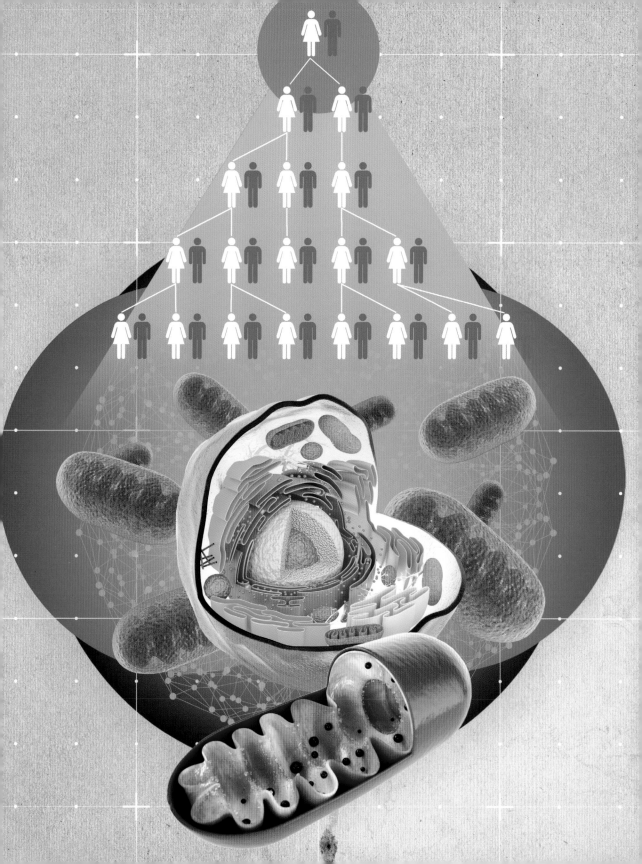

THE HUMAN Y CHROMOSOME

the 30-second theory

In many animals and plants, the sexes are genetically determined. For instance, in most of these organisms the presence of two X chromosomes drives the development of a female animal or plant. In contrast, males have an X and a Y chromosome. Y chromosomes exist in mammals, plants, and many other organisms, such as insects. The X chromosomes are generally large and full of genes, whereas the Y chromosomes are smaller and carry few genes. Although the Y chromosomes in plants and animals do not originate from a common ancestor, the logic of their evolutionary histories is the same. The X and Y chromosomes evolved from a pair of identical chromosomes through a process of differentiation linked to the appearance of the male-determining gene on the Y. Once the Y chromosome emerged, other alleles important for male reproduction accumulated around the sex-determining region. Subsequently, chromosomal rearrangements prevented the exchange of genetic material between the ancestral X and Y chromosomes. This process accelerated the evolution of the Y chromosome, which lost most of its genes and seems to be on its way to disappearing. The Y chromosome is the genetic heritage transmitted from fathers to sons for millions of years.

3-SECOND THRASH

X and Y chromosomes derive from a standard pair of chromosomes that underwent a divergence process triggered by the appearance of a male-determining gene on the Y.

3-MINUTE THOUGHT

The degeneration process of the Y chromosome has been slow. It has lost thousands of its original genes through evolution. However, most of the now essential genes on the Y are present in several back-up copies. Since the divergence of humans and chimpanzees, the human Y chromosome has not lost a single gene. This means that the Y will still be around for millions of years to tell the story of the male lineage of humankind.

RELATED TOPICS

See also
X-CHROMOSOME
INACTIVATION
page 84

SEX
page 98

3-SECOND BIOGRAPHY
CLARENCE MCCLUNG
1870–1946
American biologist who discovered the role of sex chromosomes in sex determination

30-SECOND TEXT
Reiner Veitia

You have a 50/50 chance of being born male (XY) or female (XX). Y chromosomes determine sex but do not contain genes that code for vital functions—these are found on the X chromosome.

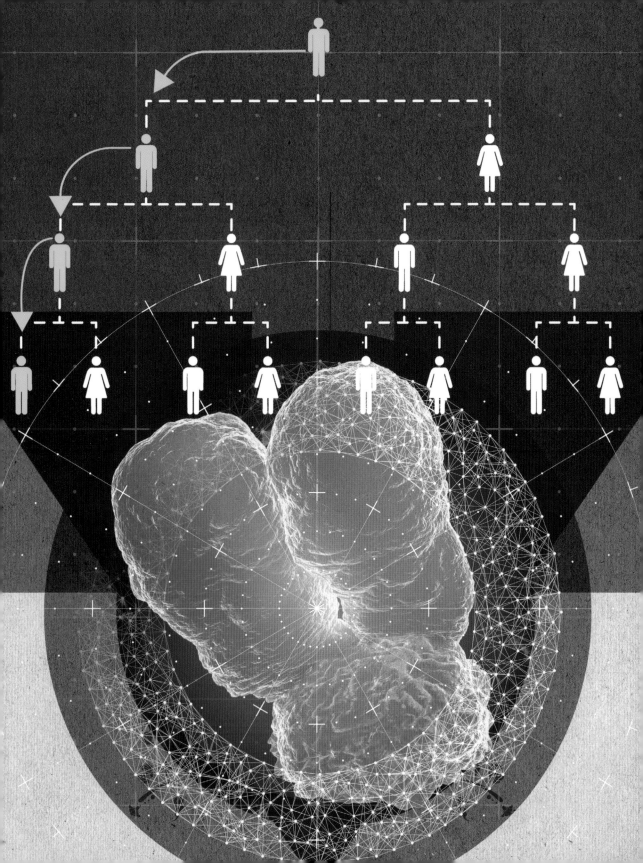

September 25, 1866
Born in Lexington,
Kentucky, United States

1886
Graduates with a
bachelor of science
degree from the State
College of Kentucky

1890
Receives his PhD from
Johns Hopkins University

1891–1904
Professor at Bryn
Mawr College

1904–28
Professor of
experimental zoology
at Columbia University

1909
Begins his pioneering
work on the fruit fly
Drosophila melanogaster

1911
Establishes the Fly
Room at Columbia

1915
Publishes his book
*The Mechanism of
Mendelian Heredity*

1919
Elected foreign member
of the Royal Society
of London

1922
Delivers the
Croonian lecture at
the Royal Society

1928–41
Professor at California
Institute of Technology

1933
Awarded the Nobel Prize
in Physiology or Medicine

December 4, 1945
Dies at the age of 79

THOMAS HUNT MORGAN

Thomas Hunt Morgan pioneered
work using the simple fruit fly *Drosophila* as
a genetic model to establish the key role of
chromosomes in inheritance. Born in 1866 in
Lexington, Kentucky, Morgan had an intriguing
ancestry: he was a nephew of Confederate
general John Hunt Morgan and the great-
grandson of Francis Scott Key, author of
the words to "The Star Spangled Banner."

Morgan showed a great interest in nature
and natural history from an early age, and
during his childhood he collected birds, birds'
eggs, and fossils. He started college at the age
of 16, receiving his bachelor's degree at the
University of Kentucky, and a PhD at Johns
Hopkins University.

From 1891 to 1904, Morgan was a professor
at Bryn Mawr College, a women's university
near Philadelphia, where he taught biology
and natural sciences. In 1904, he joined
the staff at Columbia University and there
he established the "Fly Room" to determine
how a species changed over time. Morgan was
largely responsible for establishing the fruit
fly (*Drosophila melanogaster*) as a model
experimental organism to study genetics.

Morgan carried out a particularly influential
study that confirmed the chromosome theory
of inheritance. In work leading up to his most
famous studies, Morgan engaged one of his
graduate students to rear fruit flies in the dark,
hoping to produce flies whose eyes would
atrophy from disuse and disappear in future
generations. Even after many consecutive
generations, however, the flies appeared to
have no noticeable changes despite repeated
attempts at inducing mutations by treatments
with agents such as X-rays and radium.

After two years, Morgan finally obtained an
interesting result when a true-breeding line of
Drosophila produced a male fruit fly with white
eyes rather than the normal red eyes. Morgan is
said to have carried this fly home with him in a
jar, put it by his bedside at night while he slept
and then taken it back to the laboratory during
the day. It was this white-eyed fly that allowed
him to confirm that a gene affecting eye color
in fruit flies is located on the X chromosome.
Morgan concluded that red eye color and X
(a sex factor that is present in two copies in
the female) are combined and have never
existed apart.

In 1928, he left Columbia University and
became professor of biology at the California
Institute of Technology at Pasadena (Caltech).
He established a Division of Biology there that
has produced no fewer than seven Nobel Prize
winners. In 1933, he was the first geneticist to
receive a Nobel Prize. He remained at Caltech
until his death in 1945, at the age of 79.

Robert Brooker

CENTROMERES & TELOMERES

the 30-second theory

When the cells in our bodies

divide, the chromosomes need to copy themselves. The resulting chromatid pairs are held together by a structure known as the "centromere," which assembles a complex motor that segregates chromosomes during cell division. A protein complex called the "kinetochore" attaches to the centromere and helps it pull the chromatids to opposite ends of the cell during division. In this way the chromatids end up in separate daughter cells. When a chromosome is replicated, the enzymes that duplicate DNA cannot continue to the end of the chromosome (the ends of each chromosome are called the telomeres). This poses the challenge to the cell of how to copy the complete chromosome without losing the sections at the ends. This is solved by repeating end segments, which cap the end sequence to prevent deterioration. The telomeres are replenished by the enzyme telomerase. Telomeres and telomerase play important roles in human disease. Telomere shortening is associated with ageing diseases. Telomere dysfunction or shortening can lead to genomic instability, and occurs when a tumor is developing. Telomerase can extend the lifespan of cells and is increased in cancer cells.

RELATED TOPICS
See also
CHROMOSOMES
& KARYOTYPES
page 38

CELL DIVISION
page 50

DNA DAMAGE & REPAIR
page 70

3-SECOND BIOGRAPHIES
ELIZABETH BLACKBURN
1948–
Australian-born biologist who discovered that telomeres have a specific DNA sequence

JACK SZOSTAK
1952–
British-born biologist who showed, with Blackburn, that telomeres protect ends from destruction

CAROL GREIDER
1961–
American biologist who, with Blackburn, discovered telomerase

30-SECOND TEXT
Matthew Weitzman

The center of a chromosome is known as the centromere. Its ends are the telomeres.

3-SECOND THRASH
Each chromosome has a constriction point called the centromere that helps segregate chromatids during cell division. Each also has telomeres that protect chromosome ends from deterioration.

3-MINUTE THOUGHT
When the cell divides, it is important that daughter cells have the same number of intact chromosomes. There are two chromosome structures that help to achieve this: the telomeres at the ends and the internal centromere. The telomeres prevent loss of essential genetic material from the ends, and the centromere allows the daughter strands of replicated chromosomes to segregate into daughter cells.

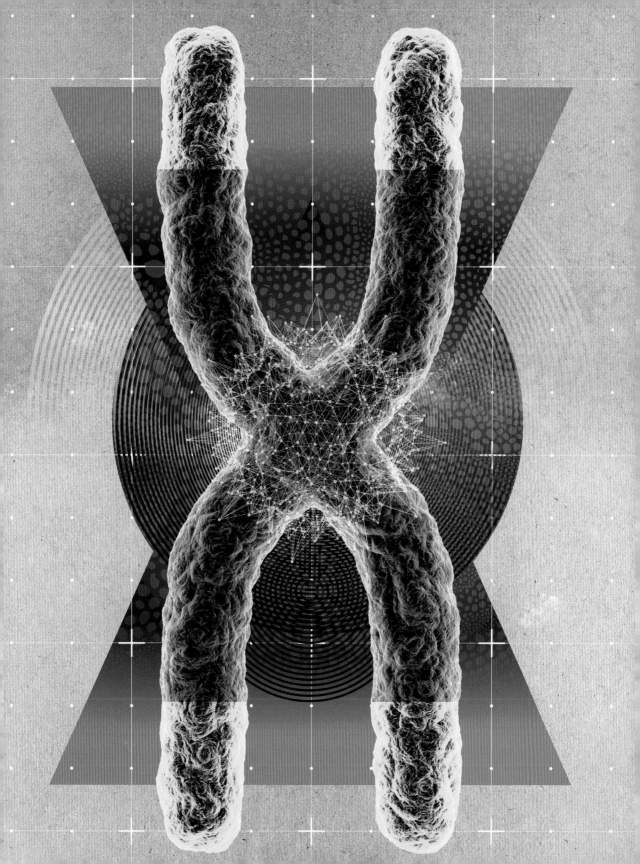

THE CELL CYCLE

the 30-second theory

3-SECOND THRASH
The four phases of the cell cycle involve cell growth and DNA replication and result in cell division to create two daughter cells.

3-MINUTE THOUGHT
The progress through the cell cycle is highly regulated to ensure that all the chromosomes are intact and that conditions are appropriate for a cell to divide. Checkpoint proteins delay the cell cycle until any problems are detected and fixed. If problems cannot be fixed, cell division is aborted. If checkpoint proteins become mutated, quality control is defective and the cell cycle can result in undesirable genetic changes that cause additional mutations or cancer.

The adult human body contains somewhere between ten trillion and 50 trillion cells. That's more than 10,000,000,000,000! It is almost an incomprehensible number. Even more amazing is the accuracy of the process that generates these cells. Virtually every cell in your body contains chromosomes that have basically identical DNA sequences, except for a few rare mutations. The cell cycle is the process by which a mother cell divides to produce two daughter cells. It is a highly regulated process in all species because it must ensure that cell division occurs at precisely the right time and without errors. The cell cycle involves cell growth, DNA replication, and cell division. It is orchestrated in four phases: G1, S, G2, and M. In the G1 phase the cell decides to divide, dependent on the presence of the proper signaling factors, growth hormones, and a sufficient supply of nutrients. During the S phase the cell copies all its genetic material and synthesizes DNA. During the G2 phase, the cell prepares itself for division. In the final M phase the cell nucleus divides, and the two daughter cells separate by a process that is known as "cytokinesis."

RELATED TOPICS
See also
THE CELL NUCLEUS
page 36

CELL DIVISION
page 50

THE GENETICS OF CANCER
page 112

3-SECOND BIOGRAPHIES
LELAND HARTWELL
1939–
American yeast biologist who identified the fundamental role of checkpoints in cell cycle control

PAUL MAXIME NURSE
1949–
English Nobel Prize-winning geneticist who identified the key proteins for the transition from one phase of the cell cycle to another

30-SECOND TEXT
Robert Brooker

The cell cycle happens in four phases, which are shown in the illustration to the right, next to the orange (G1), green (S), blue (G2), and purple (M) arrows.

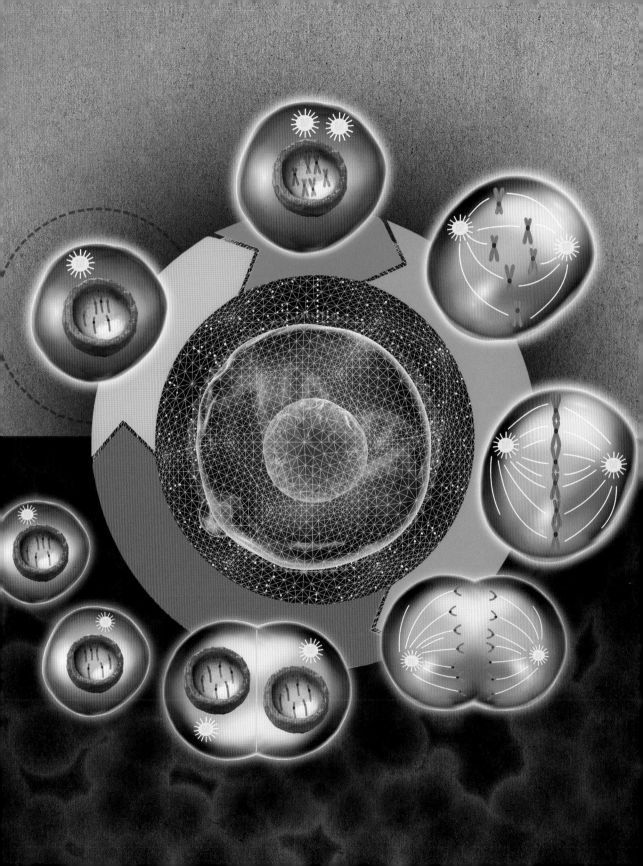

CELL DIVISION

the 30-second theory

3-SECOND THRASH

Mitosis and meiosis are cell division processes that generate the daughter cells that make up the tissues of our bodies, as well as our reproductive cells.

3-MINUTE THOUGHT

Cell division is a highly regulated process. Chromosome segregation to the daughter cells is key to the success of cell division. In cells generated by mitosis, if a chromosome lags it results in one daughter cell with only a single copy of that chromosome (instead of two) and three copies of that chromosome in the other daughter cell. Errors in mitosis can lead to cancer cells, which have unregulated cell growth.

Living organisms grow and

reproduce by cell division. Every day, cells in our bodies divide to give two daughter cells, both of which inherit the genetic material and the small cell compartments (called "organelles"), such as mitochondria, from the mother cell. There are two different types of cell division: mitosis and meiosis. Mitosis generates daughter cells that are identical to the mother cell, while meiosis produces gametes (egg and sperm cells) for sexual reproduction. Before mitosis, the cell copies all its DNA and most of its components to make sure that the daughter cells will receive the same amount of DNA and proteins. By contrast, meiosis shuffles the DNA and creates reproductive cells that contain half the amount of DNA. In humans, most cells contain 46 chromosomes, whereas the gametes contain only 23 chromosomes. When an egg and a sperm fuse to create a baby, each cell contributes half of the DNA material. When one of the gametes has an abnormal number of chromosomes (for example, an extra copy of chromosome 21), the resulting individual also has an abnormal chromosomal count (in this case known as "trisomy 21").

RELATED TOPICS

See also
THE CELL CYCLE
page 48

THE GENETICS OF CANCER
page 112

3-SECOND BIOGRAPHIES

JOHANN BÜTSCHLI
1848–1920
German zoologist who is credited with the discovery of mitosis

OSCAR HERTWIG
1849–1922
German zoologist who discovered meiosis

30-SECOND TEXT
Reiner Veitia

During mitosis, all the DNA contained in the cell is copied to create daughter cells that are exact replicas of the mother cell.

GENES & GENOMES

GENES & GENOMES
GLOSSARY

base excision repair (BER) Cellular mechanism that repairs damaged DNA throughout the cell cycle. BER removes small errors in the genome to protect against harmful mutations.

chromatin Complex formed along the DNA in eukaryotic cells. Chromatin is composed of proteins called histones as well as non-histone proteins. The structure of chromatin plays a key role in regulating gene expression.

eukaryote Organism composed of one or many cells each with a distinct nucleus and cytoplasm. There are also living cells without a nucleus, such as bacteria, called "prokaryotes."

exons and introns Messenger RNA is edited by a process called splicing that removes introns and maintains parts called exons. The exons are joined together to make the mature mRNA and this information can be used to create proteins. The genome is the complete set of genes and the complete set of exons is called the exome.

genome Complete set of genetic material in an organism or a cell. Genomics is the study of an organism's genome, focusing on its evolution, function, and structure. The genome must be very carefully monitored to make sure any errors are detected and corrected. This is referred to as "maintaining genome integrity."

genotoxic Property of chemicals that damage the genetic information within a cell by causing mutations in the DNA. Genotoxic chemicals can kill cells or cause diseases such as cancer.

genotype DNA sequence of a cell or the alleles carried by an organism that determines a specific characteristic (called a "trait" or a "phenotype") of that cell or organism.

germ cell Biological cell that gives rise to the gametes for sexual reproduction. Germ cells undergo meiosis, followed by cellular differentiation to produce mature gametes, either eggs or sperm. Gametes contain the genetic information that will be transmitted to the next generation.

mRNA (messenger RNA) Molecule that represents a copy of DNA and that contains the information to make a protein. One strand of the DNA of a gene is transcribed into a mRNA copy that is translated to produce a protein. The mRNA contains the information for encoding a functional protein.

natural selection Process through which the organisms best adapted to their environment survive and reproduce. Natural selection is a key mechanism in Charles Darwin's theory of evolution.

nucleotides Building blocks used to make DNA or RNA. Strings of nucleotides are called nucleic acids. In DNA there are four nucleotides (referred to by the letters T, C, G, and A) and in RNA there are four ribonucleotides (U, C, G, and A). Nucleotides are also called bases. DNA bases can be paired: A pairs with T, and C pairs with G.

phenotype Observable characteristics or traits of a cell or an organism (such as shape, development, biochemical or physiological features, or particular behaviors). The phenotype is influenced by the genotype within the genome.

silencing Regulation of a gene by shutting down its expression. As cells only use a fraction of their genes at any given time, the rest of their genes are repressed or silenced. Cells have mechanisms to activate or silence genes at precise times. Researchers can use these silencing mechanisms to reduce gene expression in the laboratory or even to treat disease.

somatic cells Biological cells that form the main body of an organism. There are more than 200 different types of somatic cell in the human body and these make up all the different organs and tissues. Somatic cells are not transmitted to the next generation and are distinct from germ cells and gametes.

splicing Editing of the newly transcribed messenger RNA to remove introns and paste together exons. Splicing is performed by a large protein machine called the spliceosome. Splicing is a way in which the cell can generate different proteins from the same gene by editing together different exons.

transcription Process for turning DNA genetic information into RNA. This is done by an enzyme machine called RNA polymerase that builds an RNA polymer using the DNA as a template. Transcription profiling involves measuring the amount of RNA for every gene in the cell.

transposons DNA sequence that can change its position within a genome. Transposons are sometimes called transposable elements or "jumping genes." Scientists have learned to exploit transposons—for example, the "Sleeping Beauty transposon" system is used in genome engineering.

WHAT IS A GENE?

the 30-second theory

Genes can explain part of the differences between us—whether we are tall or short, whether we have brown or blue eyes, and why we resemble our parents. Your mother gave you half of her genes, and your father half of his, so that each of us carries a completely unique collection of genes (with the exception of twins, who share identical genes). So why does a daughter have the curly hair typical of her father? Because she received the "curly hair" gene from her father and because "curly hair" is usually dominant over the recessive gene for "straight hair." Genes are detected through trait differences. They correspond to distinct DNA sequences at a given chromosome location. Research into how genes can affect visible traits led to a second definition of the word "gene": it is also a stretch of DNA that is copied into a ribonucleotide molecule or a protein, with a known function. For example, the keratin gene is used to produce the keratin protein that makes up our hair. In mice, dogs, and humans, a single mutation in the DNA sequence of the keratin gene can explain the difference between straight hair and curly hair.

3-SECOND THRASH
A gene alone is an inert DNA molecule with no effect. But changing one gene into another within an organism can produce a visible difference.

3-MINUTE THOUGHT
A human being carries as many genes in its genome as a small nematode worm. Many species (including the mouse, the pufferfish, red clover, onions, and wheat) appear to have more genes than humans do. Therefore, the complexity of life is not simply determined by the number of genes.

RELATED TOPICS
See also
DNA CARRIES THE
GENETIC INFORMATION
page 20

JUMPING GENES
page 58

GENE EXPRESSION
page 64

3-SECOND BIOGRAPHIES
WILHELM JOHANNSEN
1857–1927
Danish botanist who coined the terms "gene," "genotype," and "phenotype"

WILLIAM BATESON
1861–1926
British biologist, the first to coin the term "genetics"

THOMAS HUNT MORGAN
1866–1945
American biologist who won the Nobel Prize for his findings on genes and their location on chromosomes

30-SECOND TEXT
Virginie Courtier-Orgogozo

Your genes give you many of your physical traits, including hair color and texture.

JUMPING GENES

the 30-second theory

Transposable elements or

"jumping genes" are DNA sequences that can move to other sites in the genome. They were first described by Barbara McClintock, who observed changes in the color of corn kernels resulting from moving genes. They can move by "copy and paste" (where the original remains in its location) or "cut and paste" (where the original moves to the new location). Transposable elements, known as "transposons," make up a large fraction of the human genome. Most of the transposons are inactive, but when active they can affect the health of the genome, resulting in mutation and disease or altering how neighboring genes behave. Transposons can also drive the evolution of the genome by shuttling DNA to new locations and thereby generating genetic diversity. They have been adapted as tools for biologists to mutate and tag genes throughout the genome, enabling identification of the genes responsible for specific traits. The principle of "jumping genes" has also been harnessed to insert DNA sequences into the genome. The "Sleeping Beauty transposon" is a synthetic DNA transposon resurrected in 1997 from a fish genome; it has been used as a tool to insert specific DNA sequences into genomes of vertebrate animals during gene therapy.

3-SECOND THRASH
"Jumping genes" are sequences of DNA that can move or "jump" from one location in the genome to another.

3-MINUTE THOUGHT
Transposable elements (transposons) are DNA sequences that can change position in the genome. They make up roughly half of the human genome and are important for the workings and evolution of the genome. They can also be exploited as tools to modify the genome of cells or of a living organism.

RELATED TOPICS
See also
DNA CARRIES THE
GENETIC INFORMATION
page 20

GENE THERAPY
page 138

GENOME EDITING
page 152

3-SECOND BIOGRAPHY
BARBARA McCLINTOCK
1902–92
American cytogeneticist who discovered that genes could move from place to place on a chromosome; she received the 1983 Nobel Prize in Physiology or Medicine

30-SECOND TEXT
Matthew Weitzman

McClintock's work on transposable elements in maize wasn't fully recognized and accepted by the field until over 30 years after her initial discoveries.

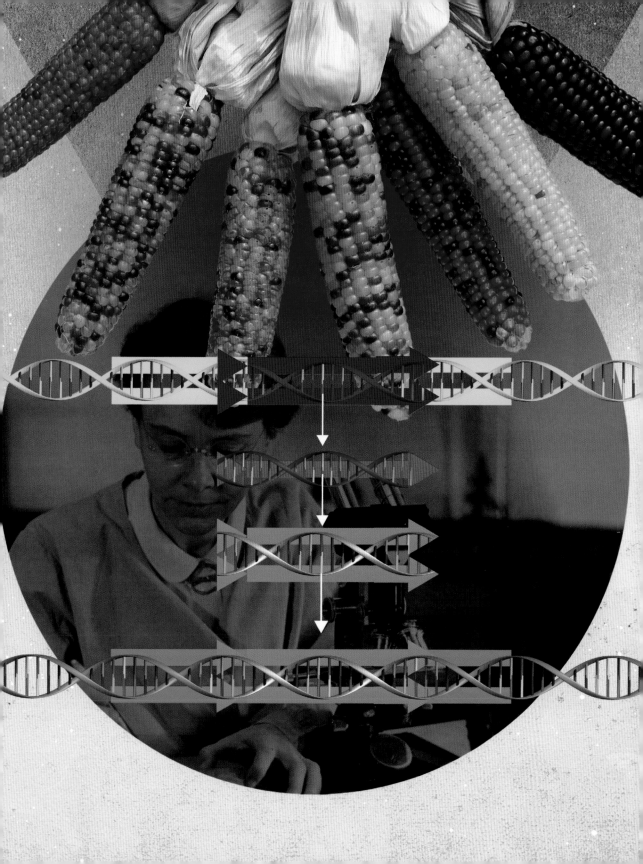

GENE SPLICING

the 30-second theory

Information coded in the DNA

sequences of genes is used to produce proteins. The first step is the transcription of the DNA sequence of a gene into a messenger RNA (mRNA) molecule. A surprising discovery several decades ago was that most of the genes of animals and plants are "split": they have parts that contain information to code for proteins and parts that do not. The protein-coding segments of genes are called exons. They are separated by long sequences that do not encode protein information, called introns. The mRNA first transcribed from a gene contains all the exon and intron sequences. But the introns are then removed by a process called gene splicing and the exons join together in the right order to create the final mRNA. One can imagine the initial mRNA as a mixture of meaningful words (exons) and gibberish (introns). Gene splicing changes the initial mRNA reading "thisiscmhazdbwthewayqtrncdbgenestalk" by removing the gibberish and joining the meaningful words together to generate the final message of the gene reading "this is the way genes talk." Alternative splicing removes different introns and joins exons to make different protein variants from the same gene. Gene splicing is an exact process that precisely removes only intron sequences from mRNA.

RELATED TOPICS
See also
THE CENTRAL DOGMA
page 28

WHAT IS A GENE?
page 56

GENE EXPRESSION
page 64

3-SECOND BIOGRAPHIES
RICHARD ROBERTS
1943–
British biochemist and molecular biologist and co-discoverer of "split genes"

PHILLIP SHARP
1944–
American molecular biologist who discovered that most genes are "split" into exon and intron segments

THOMAS CECH
1947–
American biologist who described gene splicing

30-SECOND TEXT
Mark Sanders

Gene splicing errors can play a role in genetic diseases and may lead to cancer.

GENOTYPE & PHENOTYPE

the 30-second theory

3-SECOND THRASH
The genotype of an individual determines its phenotype, through interactions with the rest of the genome and the environment.

3-MINUTE THOUGHT
The genotype (G) for a particular gene does not always lead to the same phenotype (P). This depends on the interaction of the relevant alleles with other alleles elsewhere in the genome, which can reduce or enhance the phenotype. But the environment (E) can deeply influence the expression of the genotype. This is described in the following formula:
$G + E + GxE \rightarrow P$
(G = genotype, E = environment, and GxE = their interaction).

Most organisms within a population are different from each other. These differences are mostly due to underlying genetic variations. The genotype of an individual describes its genetic make-up, be it at the single-gene or whole-genome level. Most animals can carry a maximum of two versions of each gene—or "alleles." The combination of such alleles across the genome is unique to each individual and constitutes its genetic fingerprint. Only identical twins, developing from one fertilized egg, share the same genotype. Yet even they bear differences, owing to small variations that appear after their conception. The phenotype is the set of observable or measurable characteristics of an individual, such as the color of the eyes, height, and so on. For example, in garden peas, the character white flowers (the phenotype) is determined by the genotype pp (homozygous), whereas the underlying genotype for purple flowers is PP or Pp (heterozygous). Identical variations (genotypes) in two individuals may produce the same phenotype, but this is not always so, because the phenotype is the manifestation of the interactions between the genotype and the environment.

RELATED TOPICS
See also
GENES & ENVIRONMENT
page 78

TWINS
page 92

GENETIC FINGERPRINTING
page 120

3-SECOND BIOGRAPHY
WILHELM JOHANNSEN
1857–1927
Danish botanist who coined the terms "phenotype" and "genotype" to distinguish heredity from its results

30-SECOND TEXT
Reiner Veitia

Everyone's genotype is unique and shared by nobody else. The only exception to this is identical twins, who share practically identical genotypes, although their phenotypes (physical traits) may still differ.

GENE EXPRESSION

the 30-second theory

Nearly all the cells in your body

share the same DNA, yet each cell type is equipped for a specific biological function. It turns out that not all your cells read all the genetic information in the genome at the same time. Your DNA contains all the information needed to make more than 25,000 different proteins, but each cell makes only the proteins it requires to function and will "read" just a fraction of all the genes at a given time. To make a protein, cells have to "transcribe" the DNA information into RNA and then "translate" it into the protein. Researchers say that genes are either expressed (turned "on") or repressed (turned "off"). Upstream of each gene there is a piece of DNA called the promoter, which works like a kind of switch to turn transcription on or off. There are many regulatory mechanisms to make sure that the switch is on at the right time and that each gene is expressed at the right level for that particular cell function. There are particular proteins that can recognize the switches and regulate the amount of RNA produced. The cell can also control gene expression by determining how quickly the RNA is degraded.

3-SECOND THRASH
Each cell expresses only a fraction of all the genes in the genome so that it makes the right proteins for its cellular needs.

3-MINUTE THOUGHT
Today researchers have sophisticated technologies to measure all the thousands of genes that can be expressed at the same time. By performing gene expression profiling, they can make predictions about the identity of a cell and the functions of the genes that are expressed together. Some essential genes are expressed in most cells, whereas others are expressed only in very specialized tissues.

RELATED TOPICS
See also
THE CENTRAL DOGMA
page 28

WHAT IS A GENE?
page 56

GENOTYPE & PHENOTYPE
page 62

3-SECOND BIOGRAPHIES
JACQUES MONOD
1910–76
French geneticist who worked out how genes are expressed by studying gene repression in bacteria

ROGER KORNBERG
1947–
American biochemist who pioneered work into the molecular machinery that turns genes on

30-SECOND TEXT
Jonathan Weitzman

Heat maps, such as those shown in the illustration to the right, are used to study how genes are expressed in various experiments.

June 16, 1902
Born in Hartford,
Connecticut

1918–31
Carries out her
undergraduate and
graduate work in the
College of Agriculture
at Cornell University

1933–34
Receives a Guggenheim
Fellowship to train in
Germany with geneticist
Richard B. Goldschmidt

1936–41
Serves as assistant
professor at the
University of Missouri

1941–92
Works in the Department
of Genetics at Cold
Spring Harbor, where
she discovers
transposable elements

1944
Becomes only the third
woman to be elected to
the National Academy
of Sciences and the first
female president of
the Genetics Society
of America

1970
Becomes the first woman
to be awarded the
National Medal of Science

1981
Publishes her seminal
study *The Chromosomal
Constitution of Races
of Maize*

1983
Receives the Nobel Prize
in Physiology or Medicine

1987
Publishes her book
*The Discovery and
Characterization of
Transposable Elements:
The Collected Papers
of Barbara McClintock*

September 2, 1992
Dies after a brief illness
at the age of 90

2005
Commemorated—with
John von Neumann,
Josiah Willard and
Richard Feynman—on U.S.
Postal Service *American
Scientists* postage stamps

BARBARA McCLINTOCK

Born in Hartford, Connecticut, in

1902, Barbara McClintock was an independent young woman from the start, who early in life developed what she later called a "capacity to be alone." She was eager to study science at a time when few women attended college. A keen sportswoman, she began her scientific career as a student at Cornell University, where she received her BSc in botany in 1923. Her interests quickly focused on the structure and function of the chromosomes of corn plants.

During her long scientific career, McClintock identified many unusual features of corn chromosomes. She is best known for discovering segments of DNA that can move from one site in a chromosome to another, called transposable elements. They are also called "jumping genes," because they are inherently mobile. In one of her corn strains, she noticed that a particular site in a chromosome had the strange characteristic of showing a fairly high rate of breakage. McClintock termed this a "mutable site." In one of her strains, the mutable site caused the kernels to be speckled. By studying speckled corn kernels and observing microscopic chromosomes, McClintock showed that the mutable site could move from one chromosomal site to another. It was a transposable element.

When McClintock proposed the existence of transposable elements in 1951, her theory was met with great skepticism. Some geneticists could not accept the idea that genetic material was susceptible to frequent rearrangements. They were convinced that genetic material was highly stable and had a permanent structure. Over the next decades, however, the scientific community realized that transposable elements are a widespread phenomenon.

Barbara McClintock liked to be alone and spent countless hours examining corn chromosomes under the microscope. Not only was she technically gifted, but she also had an impressive theoretical mind that challenged conventional wisdom. Much like Gregor Mendel and Charles Darwin, she was clearly well ahead of her time. McClintock was awarded the Nobel Prize in Physiology or Medicine in 1983 for discovering "mobile genetic elements," more than 30 years after her original findings. She was the first woman to be the sole winner of this prize. She also received many other honors, including the National Medal of Science from President Nixon in 1970, and was made a foreign member of the Royal Society in 1989.

She died of natural causes at the age of 90 in Huntington, New York, on September 2, 1992.

Robert Brooker

MUTATIONS & POLYMORPHISMS

the 30-second theory

3-SECOND THRASH
Mutations change the
DNA sequence. They
are one reason why the
members of a population
differ from one another
and they are required for
evolution to occur.

3-MINUTE THOUGHT
Mutations are usually
harmful to the organism
if they change the function
or the production of the
protein encoded by a
given gene. Thousands
of different human
hereditary disorders
affecting almost every
aspect of our physical
characteristics are caused
by gene mutations.
Occasionally, however,
a mutation may change
the protein product of
the gene in a way that
is beneficial. Through the
action of natural selection
on such beneficial mutants,
polymorphisms can evolve
in populations over
many generations.

All DNA molecules, whether part
of a gene or not, are subject to changes via
mutation. These changes may be small (the
addition or deletion of a single DNA base
pair or several base pairs) or large (duplication
or elimination of a chromosome segment, or
changes in the number and structure of
chromosomes). Mutations can occur in germ
cells (sperm or egg cells in humans) or in somatic
cells (those that make up all body tissues).
Mutations are rare, occurring once per million
bases in the average human cell division cycle.
They can result from spontaneous chemical
changes of DNA bases or from environmental
factors such as chemical or radiation exposure.
While rare and sometimes harmful, mutations
are also essential, as they result in the inherited
genetic variation that forms the basis of
evolutionary change. A genetic variant is called
a "mutant" when its frequency in a population
(the number of copies of the mutant in every
100 copies of the gene) is less than 1 percent.
When evolution acts to raise the frequency
of a mutant above 1 percent it is termed
a "polymorphism," meaning "many forms." The
presence of two or more polymorphic alleles
in a population is most often the result of
mutation followed by evolution that increases
the frequency of the mutant allele.

RELATED TOPICS
See also
GENOTYPE & PHENOTYPE
page 62

DNA DAMAGE & REPAIR
page 70

DOMINANT & RECESSIVE
GENETIC DISEASES
page 104

3-SECOND BIOGRAPHIES
SEWALL WRIGHT
1889–1988
British mathematical biologist
and a founder of the field of
population genetics

HERMANN MULLER
1890–1967
American biologist who
demonstrated the mutagenic
power of radiation

BRUCE AMES
1928–
American biochemist who
developed a test to determine
if a compound causes mutations

30-SECOND TEXT
Mark Sanders

*The mind-boggling
diversity of life on Earth
is the direct result of
genetic mutations.*

DNA DAMAGE & REPAIR

the 30-second theory

DNA is damaged by constant assault from inside and outside the body and cells must do everything they can to maintain the integrity of the genome. DNA can be damaged by reactive metabolites, oxidation, radiation, genotoxic chemicals, ultraviolet light, or even by the normal copying process. These areas of damaged DNA negatively affect fundamental cellular processes: they can cause mutations that change the coding genes in the genome or rearrangements that change the structural integrity of the chromosomes. Cells must recognize and repair the damaged DNA to prevent chaos in the genome. A complex apparatus constantly surveys the genome to repair any damaged DNA. Specialized proteins act as sensors to alert the cell to DNA damage. The signals then recruit enzymes that remove damaged sections of DNA. Depending on the type of damage, different sets of enzymes and repair pathways are selected. Some inherited disorders arise from mistakes in the genes that produce these enzymes. When the repair pathways are defective or switched off, genomic instability accumulates and leads to cancer. The cancer cells also become reliant on the remaining repair pathways, and this makes them vulnerable to drugs that target the intact repair pathways.

RELATED TOPICS
See also
THE CELL CYCLE
page 48

MUTATIONS
& POLYMORPHISMS
page 68

THE GENETICS OF CANCER
page 112

3-SECOND THRASH
The human genome is under constant attack and a complex apparatus monitors for DNA damage and maintains genome integrity.

3-MINUTE THOUGHT
Every day there are thousands of potentially devastating injuries to the human genome. An intricate machinery recognizes and repairs damaged areas of DNA. DNA damage that is not repaired correctly can result in mutations and instability that can lead to life-threatening diseases such as cancer, neurodegeneration, and premature aging.

3-SECOND BIOGRAPHIES
HERMANN MULLER
1890–1967
American geneticist who discovered that X-rays could mutate and kill cells

RENATO DULBECCO
1914–2012
Italian-American virologist who discovered that repair enzymes could rescue damaged DNA

TOMAS LINDAHL
1938–
Swedish-born scientist who discovered the machinery that carries out base excision repair

30-SECOND TEXT
Matthew Weitzman

Your DNA can be damaged by exposure to UV light and by smoking tobacco.

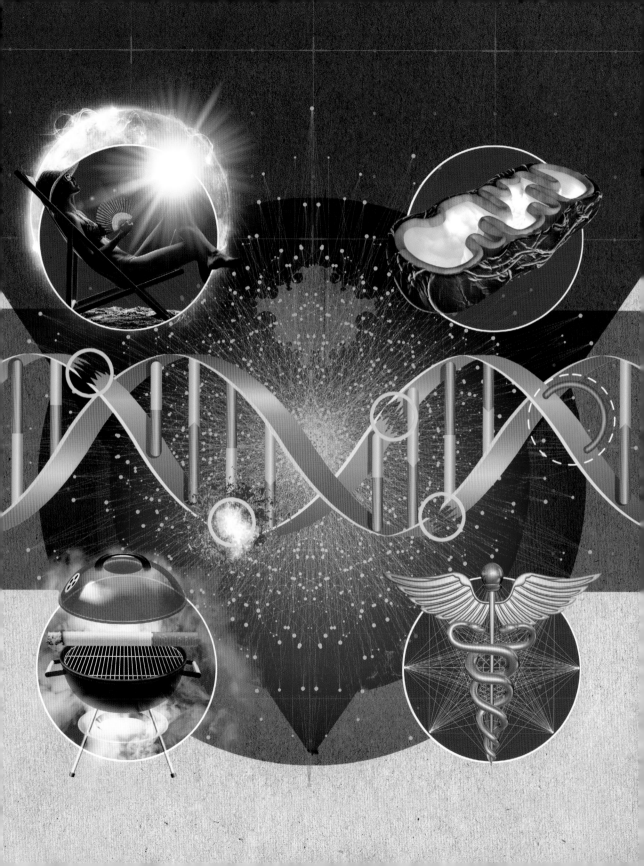

GENOME ARCHITECTURE

the 30-second theory

3-SECOND THRASH
Genomes are not randomly organized in space, but have specific architectures that allow efficient packaging of genetic material into a small volume while facilitating gene expression and other genome functions.

3-MINUTE THOUGHT
Genome architecture is dynamic: the chromosome structures are not permanent and some regions may fold and unfold over time. This is generated by proteins that bind to chromosome sequences; these include structural elements that enable long-range interactions or folding, and also regulatory elements that determine when and where genes are expressed. The organization of chromosomes in the nucleus determines how genetic information is used by the cell.

In a mammalian cell, 6ft 6in (2 meters) of DNA is packaged into a nucleus that is just a few thousandths of a millimeter wide. This packaging is not random; genomes have a specific architecture. Physical interactions within or between chromosomes play important roles in the regulation of genes, replication of DNA, and in maintaining the stability of the genome; genome architecture may be both a cause and a consequence of these functions. The packaging begins when DNA wraps around specific proteins to form chromatin. The chromatin forms a fiber that folds upon itself into loops of various sizes—from a few thousand nucleotides to large loops of several hundred thousand nucleotides. These loops are important for regulating genes, but little is known about how they form and how they affect genes. Loops are found in many organisms, including flies and bacteria. Chromosomes are also compartmentalized into different "active" or "inactive" chromatin domains. Stretches of the genome near the nuclear membrane tend to be repressed (non-active) while others in the center of the nucleus contain active genes. Biologists first defined regions of chromatin more than a century ago. Modern technologies show that the genome architecture is a scaffold that allows DNA to be correctly interpreted and copied.

RELATED TOPICS
See also
THE CELL NUCLEUS
page 36

CHROMOSOMES & KARYOTYPES
page 38

CHROMATIN & HISTONES
page 86

3-SECOND BIOGRAPHIES
CARL RABL
1853–1917
Austrian anatomist who first proposed in 1885 that chromosomes are organized into distinct regions within the nucleus

THEODOR BOVERI
1862–1915
German biologist who coined the term "chromosome territories" in 1909

30-SECOND TEXT
Edith Heard

Geneticists are still trying to define the complex influence of genome architecture on how and when genes are expressed.

EPIGENETICS

EPIGENETICS
GLOSSARY

active or silent genes Process for determining which genetic information is used in a cell at a given time. Each cell uses only a fraction of the genes for its biological functions, so the genes are either "active" (transcribed into mRNA) or "silent" (transcription is repressed and no mRNA is made).

discordance Discrepancy between genotype and phenotype that can be seen in identical twins when they have the same genetic material but may exhibit different genetic traits. Discordance in disease in identical twins can help to evaluate the influence of environmental factors.

DNA methylation Modification of DNA by the addition of methyl groups (one carbon atom with three hydrogen atoms). DNA methylation changes the function of the DNA without changing the sequence. Most DNA methylation is on the cytosine base and it often reduces gene expression.

enzyme Molecule that acts as a biological catalyst, accelerating chemical reactions in the cell. Most metabolic processes in the cell need enzymes and enzymes can modify protein functions and copy DNA. The study of enzymes is called "enzymology."

epigenetics Study of the relationship between genotype and phenotype and the investigation of effects that do not involve changes in genome sequence. The term was coined by Conrad Waddington in the 1940s to refer to "the branch of biology which studies the causal interactions between genes and their products, which bring the phenotype into being." A contemporary definition is "the study of heritable changes in genome function that occur without a change in DNA sequence."

epigenetic modifications Changes to DNA or associated proteins that affect the way the genome behaves, without directly changing the DNA sequence. These changes include DNA methylation and modifications of the histone proteins by methylation or other chemical changes. Epigenetic modifications can have a dramatic effect on gene expression and lead to repression of certain genes, called "epigenetic silencing."

epigenome The complete set of epigenetic events in an organism or a cell, including DNA methylation and histone modifications. The state of the epigenome affects the structure of chromatin and the function of the genome. Unlike the genome, which is relatively static, the epigenome can change dynamically over time and may be altered by the environment.

eukaryote Organism composed of one or many cells with a distinct nucleus and cytoplasm. There are also living cells without a nucleus, such as bacteria, called "prokaryotes."

gene dosage Number of copies of a particular gene in a genome. Most genes exist in two copies. This is not the case for some genes in males, as they have only one copy of the Y chromosome and one copy of the X chromosome. As women have two copies of the X chromosome, there is a gene dosage inequality between the sexes. Gene dosage can also be linked to disease if patients have deletions of parts of the genome or an extra copy of one chromosome (called "trisomy").

genotype and phenotype DNA sequence of a cell or an organism that determines a specific characteristic (called a "trait" or a "phenotype") of that cell or organism.

histones Family of small proteins that are associated with DNA in eukaryotic cells. Many of the histones are grouped together in balls of proteins called nucleosomes. This packaging of DNA by histones helps to organize the genome and control gene expression. The combined DNA and proteins is referred to as chromatin.

nucleosomes Basic unit of DNA packaging in eukaryotes, made up of DNA wrapped around a ball of eight histone proteins. This organization resembles beads on a string when viewed with an electron microscope.

pronucleus (pl. pronuclei) Nucleus of a sperm or an egg cell during the process of fertilization, just before they fuse. Each pronucleus carries one set of chromosomes that will combine in the nucleus of the new fertilized cell to constitute the full double set of chromosomes.

pyrimidines and purines Circular compound containing rings of two nitrogen atoms and four carbon atoms. In DNA two of the bases are pyrimidine structures: cytosine (C) and thymine (T). They pair with the other bases that are related structures called purines: guanine (G) and adenine (A).

telomere Special structure at the end of the chromosome. In eukaryotic cells the enzyme telomerase is needed to maintain telomeres at each cell division.

X and Y chromosomes Specialized chromosomes that are responsible for sex-determination. In humans, women have two XX chromosomes, whereas men have one X chromosome and one Y chromosome.

GENES & ENVIRONMENT

the 30-second theory

RELATED TOPICS
See also
WHAT IS A GENE?
page 56

GENOTYPE & PHENOTYPE
page 62

DOMINANT & RECESSIVE
GENETIC DISEASES
page 104

3-SECOND THRASH
Genes provide the genetic information to create traits (phenotypes) and the environment provides signals that affect how the program is executed.

3-MINUTE THOUGHT
The human genetic disease phenylketonuria (PKU) is an example of the interplay between genes and the environment. Most people have two functional copies of a gene that encodes the enzyme phenylalanine hydroxylase. But some people inherit two defective copies and have PKU. If PKU patients follow a standard diet containing phenylalanine in childhood, they develop severe mental impairment, defective teeth, and foul-smelling urine. But if they have a restricted phenylalanine diet they develop normally.

The environment is defined as "the conditions that surround an organism." When you plant a flower in the garden, you realize how important the environment is to its proper development: when planted in the right place and given the proper care, flowers flourish; the wrong environmental conditions, such as too much heat or cold, can have a devastating effect. The existence of every living organism, including flowering plants, is based on its genes and the environment in which it lives. Both of these factors are indispensable for life on earth. Genes provide the information to generate traits, and the environment provides nutrients and energy that can influence the traits. For example, plants have genes that encode proteins that can connect chemicals to make colorful pigments in flowers and fruits. To make such pigments, plants get chemical components from their environment—rainwater and the soil, for example. In addition, they need the right amount of sunlight, which provides the energy needed to turn the chemicals into pigments. Put simply, the environment can have a crucial influence on the way that genotype gets turned into phenotype.

3-SECOND BIOGRAPHY
ROBERT GUTHRIE
1916–95
American microbiologist who developed the neonatal heel-prick test that is used to screen newborn infants to determine if they have PKU

30-SECOND TEXT
Robert Brooker

The environment plays a vital role in an organism's survival and development. The same plant would fare very differently when planted in a warm garden compared with how it would in a hot desert.

GENOMIC IMPRINTING

the 30-second theory

RELATED TOPICS
See also
DNA METHYLATION
page 82

X-CHROMOSOME
INACTIVATION
page 84

SEX
page 98

3-SECOND BIOGRAPHIES
BRUCE CATTANACH
1932–
British geneticist who
discovered that two copies of
chromosomal regions inherited
from the same parent can lead
to abnormalities

AZIM SURANI
& DAVOR SOLTER
1945– & 1941–
Kenyan-born and Yugoslav
geneticists who discovered
that both the paternal and
maternal genomes are essential
for normal development

30-SECOND TEXT
Edith Heard

3-SECOND THRASH
Although both parents
contribute equivalent
genetic information to
the fertilized egg, the
chromosomes carry a
parental imprint and
can behave differently
depending on their
parent of origin.

3-MINUTE THOUGHT
Genomic imprinting is
found in fungi, plants, and
animals, but how and why
the parental origin of a
gene should be important
is mysterious. In mammals,
epigenetic modifications,
such as DNA methylation,
are established during
cell division and lead to
differences in expression
of imprinted genes.
Imprinting illustrates
how important appropriate
gene dosage can be.
In humans, parental
duplications of imprinted
genes can affect growth,
behavior, and create a
predisposition to cancer.

Diploid organisms are those with
two sets of chromosomes—one inherited from
the mother and one from the father. For most
genes, both parental copies are expressed
(turned on) similarly. However, in a few cases,
one copy is silent (inactive) and the other is
active depending on which parent it was
inherited from. This is known as "genomic
imprinting." It was first discovered by geneticists
in the 1970s and 1980s when they observed that
individuals with two copies of a chromosome
from the same parent had features of disease.
In the 1980s, scientists tried to create both
maternal only (gynogenetic) and paternal only
(androgenetic) diploid individuals by bringing
together two female or two male pronuclei in
mouse eggs and then transferring these eggs
into a foster mother. These eggs did not develop
normally, even when they had the same sex
chromosome complement. This was because
some genes are imprinted, with their expression
depending on the parent (maternal or paternal)
from which they were inherited. For maternal
imprinted genes, the maternal copy is silent
and the paternal copy is active, and paternal
imprinting is the opposite. We now know that
there are 100 or so imprinted genes. There are
many theories as to why imprinting evolved,
but no one yet knows for sure.

*The loss of normal
genomic imprinting can
lead to diseases such as
Prader-Willi syndrome.*

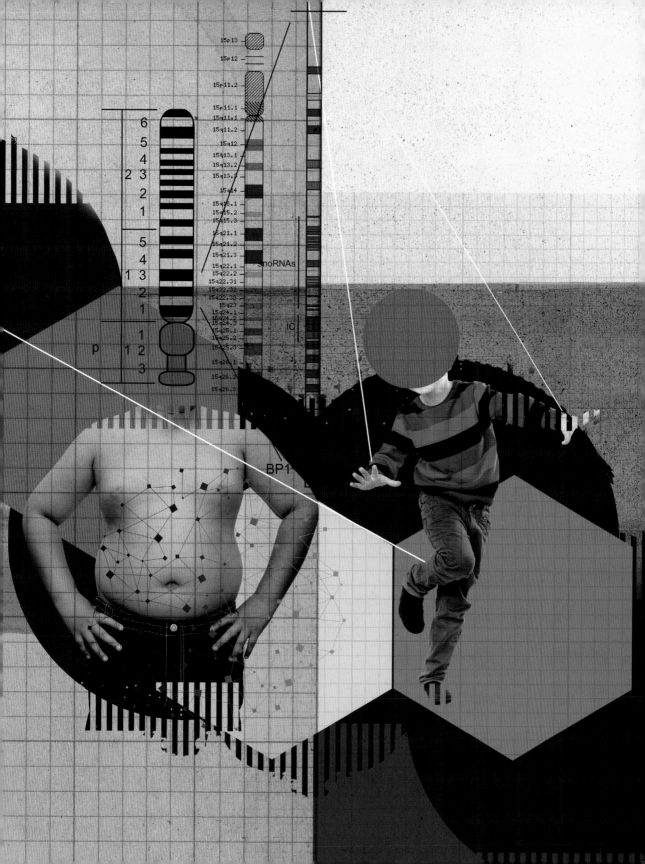

15p13
15p12
15p11.2
15p11.1
15q11.1
15q11.2
15q12
15q13.1
15q13.2
15q13.3
15q14
15q15.1
15q15.2
15q15.3
15q21.1
15q21.2
15q21.3
15q22.1
15q22.2
15q22.31
15q22.32
15q22.33
15q23
15q24.1
15q24.2
15q24.3
15q25.1
15q25.2
15q25.3
15q26.1
15q26.2
15q26.3

6
5
4
2 3
2
1

5
4
1 3
2
1

p 1 1
2
3

snoRNAs

IC

BP1

DNA METHYLATION

the 30-second theory

DNA is made up of four building blocks, called nucleotides. But one of the nucleotides, cytosine, can be modified—leading to a change in the way it is read. The modification is the addition of a CH3 methyl group to the carbon atom at the fifth position in the pyrimidine ring, creating 5-methylcytosine. This modification (methylation) changes the function of genome sequences. For example, when the promoter region of a gene is methylated it normally leads to repression (turning off the promoter switch) and less transcription of the gene. DNA methylation is essential for normal development in mammals and is critical for many epigenetic events, such as genomic imprinting and X-chromosome inactivation. DNA methylation levels may also change during the body's aging process and contribute to many types of cancer. There are enzymes that methylate DNA in specific regions and enzymes that can remove the methylation mark. Mutations in both these classes of enzymes lead to severe human diseases. And there are proteins that can specifically recognize DNA when it is methylated. There are now many techniques to identify methylated DNA in the laboratory. DNA methylation patterns are characteristic of different types of cells and different developmental histories.

RELATED TOPICS
See also
GENE EXPRESSION
page 64

GENOMIC IMPRINTING
page 80

3-SECOND BIOGRAPHIES
ROBIN HOLLIDAY
1932–2014
British molecular biologist, one of the first to suggest that DNA methylation could be an important mechanism for the control of gene expression

AZIM SURANI
1945–
Kenyan-born geneticist who discovered genomic imprinting and its link to parent-of-origin DNA methylation patterns

ANDREW PAUL FEINBERG
1970–
American scientist who discovered that changes in DNA methylation are involved in turning genes on or off in cancer cells

30-SECOND TEXT
Jonathan Weitzman

3-SECOND THRASH
Methylation is a chemical modification that changes DNA functions and provides clues about the specific characteristics and history of a given cell.

3-MINUTE THOUGHT
DNA methylation is a modification that subtly changes "letters" or nucleotides in DNA. It is a bit like accents in languages like French or Spanish. They change the way words are read and their meaning, without changing the order of the letters themselves. Just as a mistake in an accent can change the meaning of a sentence, so changes in DNA methylation can have severe consequences, leading to disease.

DNA methylation changes the way the genome functions.

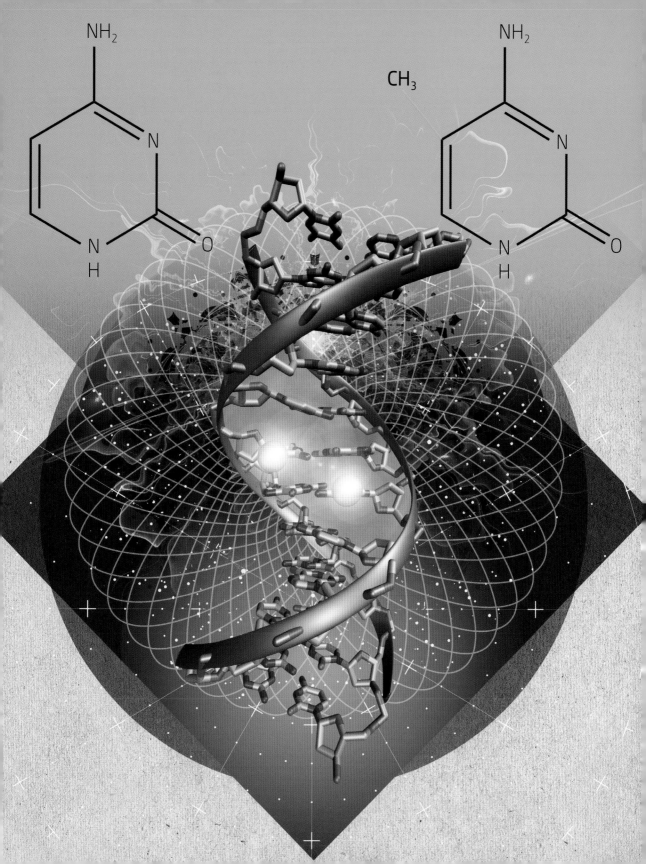

X-CHROMOSOME INACTIVATION

the 30-second theory

Chromosomes carry genes and are the basis of heredity. Having the right number of chromosomes and the right levels of gene expression is essential for life. In most mammals, females are cellular mosaics, because they use either one X chromosome or the other X in any given cell, rather than simultaneously expressing genes from both parental X chromosomes. Why? Males and females differ because of their sex chromosomes: females have two Xs, males have one X and one Y. X-chromosome inactivation in females balances out these apparent differences and equalizes the gene dosage. During development, each female cell shuts down the expression of nearly all the genes on one of its two X chromosomes. Usually, the choice of which X chromosome to inactivate (either the paternal or maternal X) is random, but once the choice is made the inactive chromosome state is maintained in all daughter cells. British geneticist Mary Lyon discovered X-chromosome inactivation when she noticed the patches of different-colored fur in female mice whose coat color gene exists in two versions (alleles) that were on the X chromosome. She proposed that cells expressed either one or the other allele, rather than both being expressed at the same time.

RELATED TOPICS
See also
CHROMOSOMES & KARYOTYPES
page 38

THE HUMAN Y CHROMOSOME
page 42

SEX
page 98

3-SECOND BIOGRAPHIES
CLARENCE ERWIN McCLUNG
1870–1946
American biologist who discovered that chromosomes play a role in sex determination from his study of locusts

MARY LYON
1961–2015
British geneticist who first proposed X-chromosome inactivation, which was designated Lyon's law in 2011

30-SECOND TEXT
Edith Heard

3-SECOND THRASH
The silencing of one X chromosome means that females have patches of cells that express the paternal X chromosome and others that express the maternal X chromosome.

3-MINUTE THOUGHT
Animals evolved different strategies to manage gene dosage inequality between sexes. In mammals, females shut down one X chromosome. In contrast, the fruit fly *Drosophila* doesn't turn off genes in females, but rather increases the expression of genes on the single X chromosome in males until they reach the levels in females. The worm *C. elegans* achieves the balance by halving the expression of genes on both X chromosomes in females.

Patches of different-colored fur are a result of the X-chromosome inactivation process.

CHROMATIN & HISTONES

the 30-second theory

3-SECOND THRASH
Eukaryotic DNA is wrapped around balls of proteins to create a highly packaged structure resembling beads on a string.

3-MINUTE THOUGHT
Histone proteins can be modified in complex ways that affect the way DNA is accessed by eukaryotic cells. Some of these modifications can predict whether genes are expressed or silenced. Researchers call the sum of all these modifications the "epigenome." Different cell types have different epigenomes and can express different genes.

The length of all the DNA in a human cell is about 6 ft 6 in (2 meters). This has to all be squeezed into a nucleus that is around ten microns (that's ten-millionths of a meter) wide. That sounds like one mighty challenge. To accomplish this, DNA must be packaged 10,000 times. Eukaryotic cells wrap their DNA around balls of proteins called nucleosomes, which look like beads on a string under an electron microscope. The nucleosomes are composed of eight proteins called histones. The DNA wraps twice around each nucleosome and makes close contact with the histone proteins. This packaged DNA is called chromatin, and its structure helps the cell to organize its DNA while also protecting against damage. But it poses a formidable problem of accessibility when the cell wants to read the DNA and regulate gene expression. By modifying the histone proteins, the cell creates some areas of the chromatin that are more accessible or "open" than others. In these open chromatin regions the genes can be expressed and in the closed-up regions the genes are often silenced. Today, researchers are mapping these chromatin regions to understand how genome organization impacts gene expression.

RELATED TOPICS
See also
GENE EXPRESSION
page 64

GENOME ARCHITECTURE
page 72

3-SECOND BIOGRAPHIES
WALTHER FLEMMING
1843–1905
German cytologist who first observed chromatin structures when he stained cells with basophilic dyes

ALBRECHT KOSSEL
1853–1927
German biochemist who discovered the proteins around which DNA is wrapped

30-SECOND TEXT
Jonathan Weitzman

An astonishing amount of DNA is packed into each of your cells. You have many millions miles of DNA in your entire body.

November 8, 1905
Born in Evesham, England

1908
Spends the first three years of his life with his parents on a tea plantation in India

1926
Graduates in geology from Sidney Sussex College at the University of Cambridge, England

1931
Works with Hans Spemann in Germany on experimental embryology

1935
Works in Thomas Hunt Morgan's fly genetics lab in California

1940
Publishes his book *Organisers and Genes*

1940
Elected fellow of the Royal Society

1947
Becomes professor of Animal Genetics at the University of Edinburgh, Scotland

1957
Publishes his book *The Strategy of the Genes* in which he describes in depth his thoughts about the epigenetic landscape

1958
Elected to the American Academy of Arts and Sciences

1960
Publishes *Behind Appearance: A Study of the Relations Between Painting and the Natural Sciences in this Century*

1968–72
Edits the four-volume work *Towards a Theoretical Biology*

1972
Founds the Centre for Human Ecology

September 26, 1975
Dies in Edinburgh, Scotland

CONRAD HAL WADDINGTON

If Gregor Mendel is considered the father of genetics for his pioneering work defining the laws of heredity, then the father of epigenetics is undoubtedly Conrad Hal Waddington. His early training in embryology fed his interest in how organisms develop from a single fertilized cell into the complex form of the mammalian embryo. From friends and colleagues he learned about the ideas emerging in genetics at a time when the molecular features of genes were still uncertain. He performed early experiments on frogs and fruit flies trying to understand developmental biology. But he is best known for coining the word "epigenetics" in the 1940s to define the "branch of biology which studies the causal interactions between genes and their products which bring the phenotype into being." He wanted this new field to create an intersection between classic embryology, modern genetics, and evolutionary theory.

Known affectionately as "Wad" to friends and "Con" to his family, Waddington moved with ease from one discipline to another. He befriended geneticists such as Gregory Bateson and Theodosius Dobzhansky, as well as philosophers and contemporary artists like Henry Moore and John Piper. He was a prolific writer, publishing a series of books in which he proposed new concepts, invented new words (such as "epigenotype" and "cheode"), and refined his ideas about developmental biology.

But his greatest legacy is probably his concept of the epigenetic landscape. This is captured by a painting in one of his 1940s books in which the visual metaphor of a landscape is applied to the development of an embryo: at the top of a mountain sits a single ball representing the fertilized egg, the multipotent stem cell. He proposed that as the cell descends the mountain its developmental potential becomes more and more restricted—its cellular identity is fixed (he used the word "canalized") by the path it takes and the valleys it enters. He added another painting that proposed a view behind the landscape in which a series of interconnected pegs and guy ropes (representing genes) determined the terrain.

More than half a century later, researchers are now defining the molecular details that control epigenetic events. As with so many visionary concepts in biology, it may be years before we understand the details of the mechanisms that explain epigenetic mysteries.

Jonathan Weitzman

NON-CODING RNA

the 30-second theory

RNA studies have been full of surprises. When Francis Crick proposed the central dogma of molecular biology to explain protein synthesis, he positioned RNA as a messenger (called mRNA) that was important for translating genetic information in the DNA into proteins. But in recent years we have discovered many groups of RNA molecules with several roles other than that of messenger. Indeed, the vast majority of human RNA molecules (maybe as much as 98 percent) do not contain information for coding proteins, but are called non-coding RNA (ncRNA). So what do all these non-coding RNAs do? They seem to be important for fine regulation of the expression and the function of the coding RNA. For example, small RNAs called tRNA are important for translating the mRNA information and ribosomal RNAs are part of the big machine that makes proteins. Non-coding RNAs can be very short, like the RNAi molecules that interfere with normal gene functions. Or they can be very long, like the *Xist* molecule that can inactive the whole X chromosome in females or the RNAs that help cells to maintain their telomeres. All organisms, from simple yeast to humans, have evolved clever ways to use RNA molecules to regulate their genomes. Non-coding RNAs have also been linked to many diseases, such as cancer and autism.

RELATED TOPICS
See also
THE CENTRAL DOGMA
page 28

CENTROMERES & TELOMERES
page 46

X-CHROMOSOME
INACTIVATION
page 84

3-SECOND BIOGRAPHIES
CARL RICHARD WOESE
1928–2012
American microbiologist who proposed the "RNA World" hypothesis in 1967

SHIRLEY M. TILGHMAN
1946–
American molecular biologist who identified mysterious long non-coding RNA, termed "H19"

CRAIG C. MELLO &
ANDREW Z. FIRE
1960– & 1959–
American biologists who discovered RNA interference

30-SECOND TEXT
Jonathan Weitzman

Woese hypothesized that all life on earth today descends from RNA-based lifeforms.

3-SECOND THRASH
RNA does a lot more than just copy DNA sequences in order to make proteins. In fact, most RNA molecules regulate genome function, rather than code for proteins.

3-MINUTE THOUGHT
RNA molecules can do so many things and seem much more versatile than DNA. This led to the "RNA World" hypothesis, which proposes that RNA appeared on Earth before DNA and proteins, and that RNA is the origin of life on Earth. Today, researchers exploit these non-coding functions of RNA to make new experimental tools and develop novel ways to treat disease.

TWINS

the 30-second theory

Twins have mystified people for centuries—think of the biblical brothers Jacob and Esau who were said to have battled in the womb, or Romulus and Remus, mythical founders of Rome. Having two identical persons challenges what makes us unique or different. Identical monozygotic (MZ) twins are always of the same sex, whereas fraternal dizygotic (DZ) twins are genetically no more similar than siblings. MZ twinning accounts for only about 0.3 percent of pregnancies. But the frequency of DZ twins is variable, and is influenced by diet, maternal age, and fertility treatments. There is a genetically linked tendency to release two eggs to produce DZ twins, but no evidence for genetic MZ twinning. If the fertilized egg divides later in development, it can lead to conjoined twins that share body parts as well as genomes. MZ twins are natural "clones," with virtually identical genetic inheritance, so any differences (known as "discordance") point towards environmental factors. Studies of MZ twins separated at birth help researchers to find non-genetic causes of behavior or disease, as many human traits have a strong genetic influence. Studies have also shown that MZ twins become increasingly different with age, and environmental influences could explain some disease discordance.

RELATED TOPICS
See also
GENOTYPE & PHENOTYPE
page 62

GENES & ENVIRONMENT
page 78

BEHAVIORAL GENETICS
page 102

3-SECOND BIOGRAPHIES
CHANG & ENG BUNKER
1811–74
Thai-American conjoined brothers who were the original "Siamese twins"

FRANCIS GALTON
1822–1911
British scientist who pioneered twin studies and coined the term "nature versus nurture"

30-SECOND TEXT
Jonathan Weitzman

Identical twins have provided opportunities for geneticists to study the influence of the environment on the human body. Non-identical twins are as alike as siblings.

3-SECOND THRASH
Monozygotic twins have nearly identical genomes, offering a powerful tool to study the impact of the environment versus genetic effects on human traits.

3-MINUTE THOUGHT
Twins can be either identical (monozygotic, MZ) coming from a single fertilized egg that splits into two, or fraternal (non-identical or dizygotic, DZ) from two fertilized eggs, in which case they share only 50 percent of their genes. By studying pairs of MZ and DZ twins, researchers can distinguish inherited genetic effects from environmental influences and can find non-genetic determinants of diseases.

HEALTH & DISEASE

alkaptonuria Rare inherited genetic disease in which the body cannot process the amino acids phenylalanine and tyrosine. The disease is caused by a mutation in the gene for an enzyme called HGD. If the child inherits two mutant copies, one from each parent, chemicals (alkapton) accumulate in the urine, turning it a dark color that can be detected at birth.

autism Neurodevelopmental disorder involving difficulties in social interaction, communication, and behavior. Children are normally diagnosed before the age of three years old. Asperger syndrome is a milder form, with normal language and intelligence.

autoimmunity A phenomenon in which the immune system of an organism acts against its own healthy cells and tissues. Diseases caused by aberrant immune response are called autoimmune disease. Examples include celiac disease and type one diabetes.

autosome Chromosome that is not one of the sex chromosomes (X or Y). Autosomes exist in pairs, each carrying the same genes. Autosomal dominant diseases are inherited when one copy of a gene on an autosome is mutated. Autosomal recessive diseases, however, only occur when both copies of the gene have mutations. If the two copies of a given gene are different the child is heterozygote; if they are the same the child is homozygote.

brain synapses Critical functional elements of the brain. Synapses are the points of communication between the brain cells (called neurons). The brain contains billions of neurons and each is connected by synapses to thousands of other neurons. The human brain may contain as many as 100 trillion synapses. Some synapses excite the neighboring cell, while others can be inhibitory. Changes in synapses are important for the brain's capacity to learn and remember.

circadian rhythm Biological process that maintains daily oscillations of about 24 hours. The 24-hour rhythm is set by an internal biological clock that is influenced by environmental conditions.

hemoglobin Protein containing iron that is responsible for carrying oxygen around the body in red blood cells. Hemoglobin carries oxygen from the lungs or the gills to the tissues of the body. Mutations in the hemoglobin gene cause diseases such as sickle cell disease and thalassemia.

Hox genes Group of similar genes that control the body plan of an embryo from head to tail. The *Hox* proteins determine the segment structure in the embryo, such as legs, wings in flies, and vertebrae in humans. Mutations in the *Hox* genes can create body parts and limbs in the wrong places along the body. In many animals, the organization of the *Hox* genes along the chromosome is the same as the order of their expression along the length of the developing embryo (called collinearity).

HPV and cervical cancer Human papillomavirus (HPV) is a virus that has been linked to cervical cancer and genital warts. It is typically sexually transmitted. HPV is one of the most important infectious causes of cancer and may contribute to 5 percent of diagnoses. In HPV-induced cancers the virus DNA can integrate into the DNA of the host cell, wreaking havoc with the mechanisms that control normal cell growth and division.

immunity Body's biological defences to fight infection and disease. The immune system involves two components: the innate and the adaptive. The former recognizes foreign substances and reacts, while the latter involves the system of lymphocyte cells that eliminate pathogens.

immunodeficiency State in which the immune system cannot manage to fight infections. It can be caused by extrinsic factors, including viral infection or poor nutrition, but some people are born with intrinsic defects in their immune system—making them susceptible to infections. Severe combined immunodeficiency is an extreme case in which both and T and B lymphocytes are affected.

lymphocytes White blood cells in the vertebrate immune system. Lymphocytes come in different types, including natural killer cells (NK cells, which kill foreign and cancer cells), T cells (which can also kill), and B cells (which make antibodies).

Online Mendelian Index of Man (OMIM) Catalog of human genes and genetic disorders and traits, with free information on Mendelian diseases and over 15,000 genes. It is particularly focused on the relationship between phenotype and genotype.

single-nucleotide polymorphisms (SNPs) Variation in a single nucleotide at a specific position in the genome, where each variation is present to some degree within a population. SNPs cause many diseases, especially if the two variants affect protein structure and function.

SEX

the 30-second theory

Sex occurs in many different organisms, ranging from bacteria to plants and animals. Most species exist in two alternative forms known as "sexes." Even in bacteria the idea of sex is similar to what is observed in more complex organisms. Organisms of different sexes produce the gametes: males produce sperm in animals or pollen in plants, females produce the ova or eggs. These are the carriers of genetic information from each parent to their offspring. Maternal and paternal gametes fuse to produce a cell that will develop into a full organism. To keep the number of chromosomes constant within a species, each gamete must carry half the amount of DNA of the organism. The process that reduces the amount of DNA is called meiosis. The sex of an organism is often genetically determined. However, there are many organisms in which sex is determined by environmental conditions. In other cases, the same organism can be first male and then female and vice-versa. And some can even be male and female at the same time. The latter are called hermaphrodites. Most female mammals, typically have two X chromosomes whereas males have XY chromosomes. The Y chromosome carries the gene that triggers maleness. In other animals, such as birds, females have the ZW chromosomes and males are ZZ.

3-MINUTE THOUGHT
The separation of the sexes may have evolved from hermaphroditic organisms able to produce male and female gametes. This separation led to a specialization of the sexes to produce only one type of gametes. The driving force of sexual reproduction is that it helps spread combinations of advantageous mutations through the blending of the genetic material from both parents. This helps organisms face a changing environment.

RELATED TOPICS
See also
THE HUMAN Y CHROMOSOME
page 42

CELL DIVISION
page 50

X-CHROMOSOME
INACTIVATION
page 84

3-SECOND BIOGRAPHIES
AUGUST WEISMANN
1834–1914
German evolutionary biologist who proposed in 1889 the evolution of sex to create variation among siblings

CLARENCE ERWIN McCLUNG
1870–1946
American biologist who discovered the role that chromosomes play in sex determination

30-SECOND TEXT
Reiner Veitia

In humans, sex is determined as female if we are born with two X chromosomes or male if we have an X and a Y chromosome.

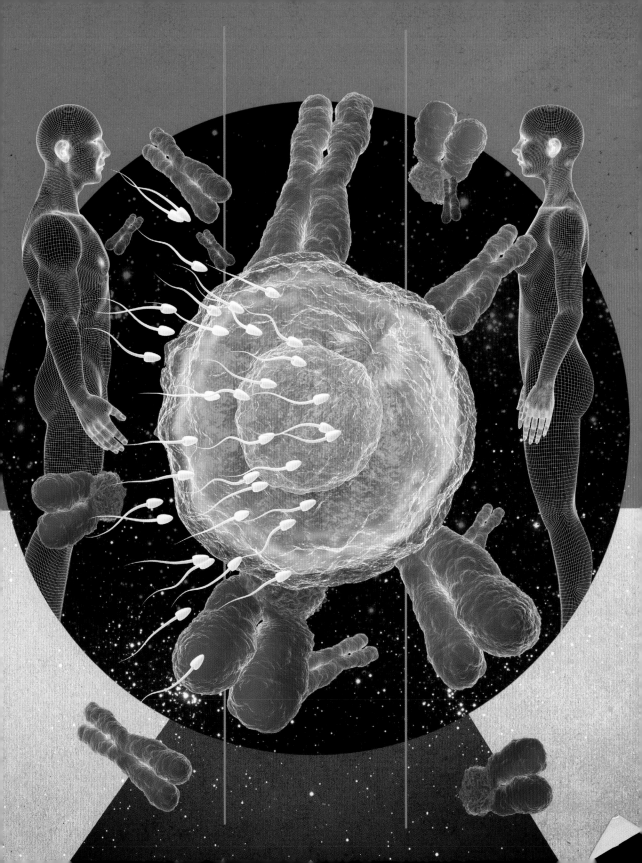

DEVELOPMENTAL GENETICS

the 30-second theory

How does a simple egg cell

transform into a complex organism made of diverse cells such as neurons, blood cells, or skin cells? Developmental biologists have devoted a great deal of attention to the problem and discovered that part of the answer lies in the genes within our genome. The gene pool of an organism is set up at fertilization and does not change over time (with rare exceptions). Therefore, all the cells in our body carry the same set of genes. But, how can cells be so different? Differences arise from variations in gene activity: for example, the hemoglobin gene is switched on in red blood cells, while photoreceptor genes are switched on in eye neurons. Developmental genetics is the study of how genes are turned on and off to control the growth and development of an organism. In developing mammal embryos a gene called *HOXD3* is switched on in a special group of cells, turning them into neck cells. Some of these cells will become neurons and others muscles or vertebrae, depending on which other genes they switch on or off. Whether a cell turns on a gene, or not, depends on its position, its internal state and the external signals it receives.

RELATED TOPICS
See also
DARWIN & THE
ORIGIN OF SPECIES
page 18

SEX
page 98

BEHAVIORAL GENETICS
page 102

3-SECOND BIOGRAPHIES
CHRISTIANE
NÜSSLEIN-VOLHARD
1942–
German biologist who, with Edward Lewis and Eric Wieschaus, won the 1995 Nobel Prize in Physiology for her work on genes controlling development of the fruit fly

SEAN B. CARROLL
1960–
American biologist who argues that morphological evolution mostly arises through changes in gene expression

30-SECOND TEXT
Virginie Courtier-Orgogozo

During development, the right genes must be switched on or off for the various cells and organs to form.

3-SECOND THRASH
As organisms develop, they switch the correct genes on or off at the right time and in the right place.

3-MINUTE THOUGHT
Certain genes (called the *Hox* genes) are expressed in stripes spanning the body from head to tail and determine the various body parts from anterior to posterior. Amazingly, these same genes are important for specifying corresponding body parts in humans, mice, and flies.

BEHAVIORAL GENETICS

the 30-second theory

RELATED TOPICS
See also
MUTATIONS & POLYMORPHISMS
page 68

GENES & ENVIRONMENT
page 78

TWINS
page 92

3-SECOND THRASH
Genetic variation influences behavioral variation, but the levels and mechanisms of genetic involvement in behavior variation in humans are largely unknown.

3-MINUTE THOUGHT
Genome-wide association studies (GWAS) seek to establish statistically significant associations between a variant and a trait. They can identify regions of the genome that contain genetic variants influencing human behavior. For complex behavioral abnormalities like schizophrenia, hundreds of variants are associated with the condition. Identifying the genes near the variants that influence human behavior is the next and most difficult challenge.

Studies in fruit flies were the first to reveal the effects of genetic variation on behavior. Genetic mutations produce proteins with abnormal functions that disrupt the development of normal behavioral responses. For example, studies of the circadian cycle identified mutations that altered the functions of the daily biological clock. Researchers also found mutations that disrupt brain synapses and affect learning and memory. Mutations in fruit flies have even been linked to courtship and mating behaviors. Behavioral genetic studies in humans are particularly challenging because of the many environmental factors affecting human behavior. Studies comparing identical and non-identical twins can indicate possible genetic influences. Studies determine the concordance of twin pairs—that is, the percentage of pairs in which both twins share a trait. Higher concordance in identical twins compared to non-identical twins suggests a genetic influence. Studies of concordance for autism, depression, and schizophrenia found 30–70 percent among identical twins compared to 5–15 percent for non-identical twins. These results imply moderately high levels of genetic influence. This effect could involve many genes, with the influence of each being relatively small.

3-SECOND BIOGRAPHIES
FRANCIS GALTON
1822–1911
English intellectual whose ideas about heredity of success led to the now discredited eugenics movement

LEE EHRMAN
1935–
American geneticist who described the relationship between genotype and reproductive success in fruit flies, paving the way for research into the genetics of behavior

30-SECOND TEXT
Mark Sanders

How individual genes contribute to complex behavioral traits is largely a mystery.

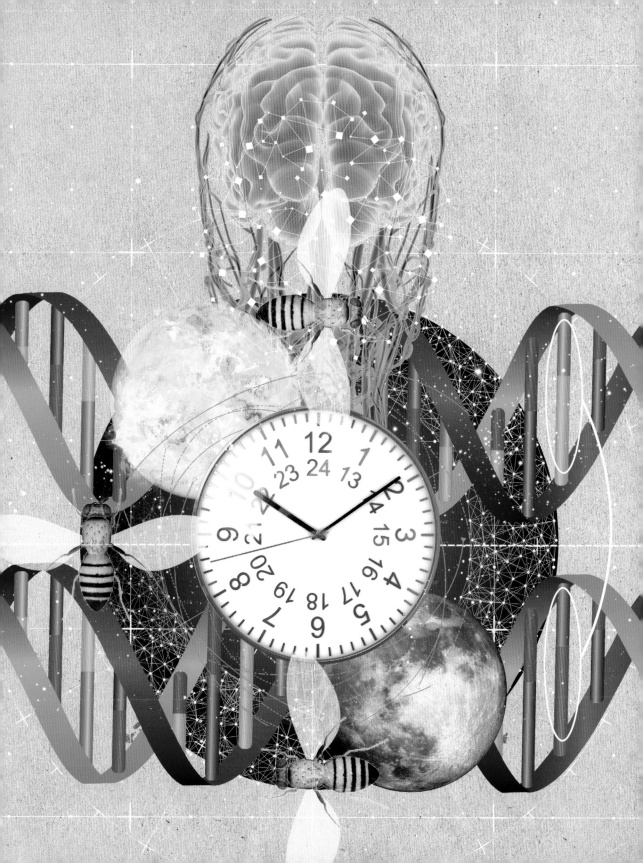

DOMINANT & RECESSIVE GENETIC DISEASES

the 30-second theory

3-SECOND THRASH

Genetic diseases caused by mutations of single genes on autosomal chromosomes or X chromosomes are inherited following the laws of heredity described by Gregor Mendel and by later researchers.

3-MINUTE THOUGHT

Mutations in human genes that are responsible for diseases are relatively rare. When mutant allele frequencies are polymorphic, it is likely that natural selection is responsible for their frequencies. For example, the life-threatening condition sickle cell disease occurs in the homozygous *aa* genotype. In malaria-rich environments, however, heterozygotes (Aa) better resist the disease than do AA homozygotes. Natural selection maintains both alleles *A* and *a* in populations.

More than 10,000 human genetic diseases are caused by mutations of single genes. These diseases are inherited as "Mendelian" traits because their inheritance follows the laws of heredity first described by Gregor Mendel. Many genetic diseases are inherited through mutations of genes carried on autosomal chromosomes—the chromosome pairs numbered one to 22 in humans. Autosomal genes can have either of two homozygous genotypes (for example, AA and aa) or the heterozygous genotype (Aa). Autosomal genetic diseases are inherited as dominant traits when one mutant allele is enough to have the disease. Diseases are autosomal recessive when two mutant copies of an allele are needed. Some genetic diseases are also caused by mutations of genes on X chromosomes. Females have two Xs and can have either of the homozygous genotypes or the heterozygous genotype. X-linked dominant genetic diseases are caused by a mutant allele on either X chromosome. X-linked recessive diseases require two mutant alleles on both female X chromosomes. In contrast, males have just one X, which expresses the trait corresponding to the X-linked allele he carries. Thus, regardless of whether the X-linked mutant allele is recessive or dominant, a male who carries the mutant allele has the disease.

RELATED TOPICS

See also
MENDEL'S LAWS OF HEREDITY
page 16

DNA CARRIES THE GENETIC INFORMATION
page 20

MUTATIONS & POLYMORPHISMS
page 68

3-SECOND BIOGRAPHIES

THOMAS HUNT MORGAN
1866–1945
American geneticist who described inheritance of genes on the X chromosome based on Mendel's laws of heredity

VICTOR McKUSICK
1921–2008
American physician and human geneticist who established a catalog of human genetic diseases that has become the Online Mendelian Index of Man (OMIM)

30-SECOND TEXT

Mark Sanders

Genetic mutations can cause many thousands of different diseases.

1857
Born in London, England

1880
Graduates from the University of Oxford in natural science

1885
Obtains a bachelor of medicine degree from the University of Oxford

1899
Appointed physician at the Hospital for Sick Children at Great Ormond Street

1902
Publishes *The Incidence of Alkaptonuria: A Study in Chemical Individuality*

1908
Delivers the Croonian Lectures on "Inborn Errors of Metabolism" to the Royal College of Physicians in London

1910
Elected fellow of the Royal Society

1914–18
During the First World War, serves as consulting physician to the Mediterranean forces (until 1919)

1918
Appointed Knight Commander

1920
Named Regius professor of medicine at the University of Oxford

1926–28
Serves as vice-president of the Royal Society

1935
Awarded the Gold Medal of the Royal Society of Medicine

1936
Dies of a heart attack in Cambridge, England

ARCHIBALD GARROD

Archibald Garrod was a born investigator. He was destined for biomedical science from the start: his older brother Alfred Henry was a zoologist and his father was the eminent doctor Alfred Baring Garrod, who had discovered the link between uric acid metabolism and gout, and pioneered rheumatoid arthritis research. After studying medicine at the University of Oxford, Archibald began his medical research exploring rare disorders such as alkaptonuria, a congenital condition in which the color of urine turns black. He collected urine and family histories from his patients. Influenced by his colleague William Bateson and the emerging new understanding of Mendelian inheritance, Garrod came to the hypothesis that metabolic variations that he called "chemical individualities" could explain such rare conditions. In 1902, he published his findings in a book, *The Incidence of Alkaptonuria: A Study in Chemical Individuality*, which presented the first case of recessive inheritance in humans. This new concept led to the identification of inherited metabolic error and started the whole field of medical genetics.

In 1908, Garrod delivered his Croonian Lectures to the Royal College of Physicians on "Inborn Errors of Metabolism." These are considered a landmark in the history of biochemistry, genetics, and medicine. He used Gregor Mendel's law of gene segregation to explain the transmission of human traits and diseases such alkaptonuria, albinism, cystinuria, and pentosuria (sometimes referred to collectively as "Garrod's tetrad"). This new concept did not immediately attract the attention of physicians, who were not interested by rare inherited traits, and geneticists, who were divided between biometricians and Mendelians.

Garrod advocated taking results from basic research into account in medicine. Together with Sir William Osler, a physician and medical educator, they were at the origin of the Association of the Physicians of Great Britain. One of the objectives of this association was to facilitate the publishing of a new type of medical journal, to record basic research that had no direct clinical application.

The advances in medical genetics owe more to Garrod's curiosity as an investigator than to his sensitivity as a physician. It was said that his bedside manner was limited to his interest in his patients' urine samples. But this combination of careful family and sample analysis has done a great deal for our understanding of genetic diseases. Today, the causes of more than 4,800 Mendelian traits have been identified in humans. Thanks in part to Archibald Garrod, medical doctors together with basic genetics researchers are now exploring new treatments for these genetic diseases by exploring how "chemical individualities" are influenced by the interaction of genetics, epigenetics, and environmental factors.

Thomas Bourgeron

GENES & IMMUNODEFICIENCY

the 30-second theory

3-SECOND THRASH
Although they are rare, there are many monogenic diseases of the immune system—ones controlled by a single gene. These can predispose people to infections, autoimmune and inflammatory diseases, allergies, and cancer.

3-MINUTE THOUGHT
Research into the molecular defects causing immunodeficiency can also teach us how the immune system works, and highlights the components of the immune system that protect us from infections. One example is the enzyme Activation Induced Deaminase (AID), which is required for making antibodies. Another is the discovery that the genetic defects that impair interferon γ production also cause selective predisposition to mycobacterial infections.

Geneticists have discovered

300 genes that are mutated in diseases of the immune system. The incidence of such mutations is 1 in 3,000–4,000 live births. A high proportion of individuals with mutations show clinical symptoms of immunodeficiency, highlighting the key roles of these genes in immune function. Immunodeficiency disorders predispose patients to infections, autoimmunity (when the body launches immune response against its own healthy cells and tissues), inflammation, allergies, and cancer. Some patients suffer from a broad susceptibility to many different microorganisms, as in severe combined immunodeficiency (SCID), while others have a surprisingly narrow spectrum of infectious susceptibility. All aspects of the immune response can be affected, including both innate and adaptive immunity, although immunodeficiency diseases frequently affect the latter. The most common genetic defects affect the production of antibodies by B lymphocytes, with the next most common involving T lymphocyte and phagocytic cell deficiencies. Early diagnosis of these disorders is important for appropriate treatment, which may include protein replacement, cell replacement, gene therapy, and targeted modulation of inflammation and autoimmunity.

RELATED TOPICS
See also
DOMINANT & RECESSIVE
GENETIC DISEASES
page 104

THE GENETICS OF CANCER
page 112

3-SECOND BIOGRAPHIES
ROBERT ANDERSON ALDRICH
1917–98
American pediatrician who showed that an immunodeficiency syndrome first identified in 1937 by Alfred Wiskott was passed from generation to generation through the X chromosome. The disease is known as "Wiskott-Aldrich syndrome"

ROBERT A. GOOD
1922–2003
American physician who is regarded as a founder of modern immunology and led the team that performed the first successful bone marrow transplant in 1968

30-SECOND TEXT
Alain Fischer

Genetic defects can affect several cell types of the immune system.

THE GENETICS OF AUTISM

the 30-second theory

Autism affects more than

1 percent of the world's population. People with autism have atypical social and communication skills and restricted interests—and they exhibit repetitive behavior. Autism is not a discrete condition but rather a spectrum of behaviors. Autism rarely emerges in isolation; it usually coexists with other psychiatric and medical conditions, including intellectual disability, epilepsy, sleep disorders, and gastrointestinal problems. More than 100 risk genes for autism have been identified. For some individuals, a single mutation is enough to develop autism (especially in individuals with both autism and intellectual disability). In contrast, the genetic architecture is more complex in some individuals, involving more than 1,000 genetic variations, each with a low effect, which additively increase the risk of autism. Many of the risk genes are key regulators of brain connectivity regulating the contact between neurons (the synapses). Changes in any of these proteins can increase or decrease the number and strength of the brain's synapses and, ultimately, connections within the brain. Current research on autism explores the role of these genes during brain development. This knowledge should improve diagnosis, care, and integration into society for individuals with autism.

3-SECOND THRASH

The genetics of autism differs from one individual to another, but most of the autism-risk genes regulate brain connectivity.

3-MINUTE THOUGHT

People affected by autism range from individuals with no verbal language skills to those with Asperger syndrome who have high cognitive functions. Most of our knowledge on the genetics of autism comes from studies of apparently monogenic forms of autism (those controlled by a single gene). Mice with these mutations display atypical social interaction and ultrasonic vocalization. Neurobiological studies show that synaptic plasticity—the property of synapses to respond to environmental stimuli—is different in people with autism.

RELATED TOPICS

See also
DOMINANT & RECESSIVE
GENETIC DISEASES
page 104

GENES &
IMMUNODEFICIENCY
page 108

3-SECOND BIOGRAPHIES

LEO KANNER
1894–1981
Austrian-American psychiatrist and physician who published the first cases of patients with autism

HANS ASPERGER
1906–80
Austrian pediatrician who reported the first patients with Asperger syndrome

30-SECOND TEXT

Thomas Bourgeron

Autism is a complex spectrum condition that can involve many genes. Often these genes are responsible for managing the connections between synapses in the brain.

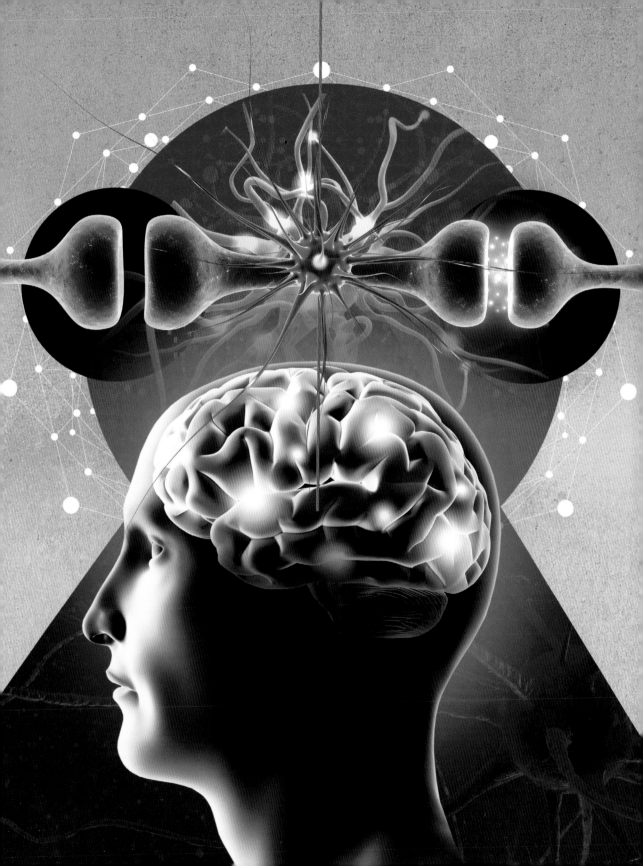

THE GENETICS OF CANCER

the 30-second theory

Cancer is a terrible disease and one of the leading causes of death worldwide. It occurs when normal cells of the body lose control of their cell cycle, so that they divide without stopping and spread around the body. Rapidly dividing cancer cells can form growths called tumors; benign tumors do not spread, whereas malignant tumor cells can invade other tissues (a process called metastasis). Cancer is a genetic disease that results from changes in the genes that normally control how cells grow and divide. There are hereditary cancer syndromes, called germline mutations, in which genetic changes are inherited from parents and can be passed on to children. But most cancers result from genetic changes that occur over one's lifetime. These are called somatic mutations and can be due to errors during cell division or mutations caused by exposure to chemical substances (such as tobacco smoke) or radiation (such as UV rays). Cancer mutations can activate genes that push cells to divide, called "oncogenes." Alternatively the mutations can inactivate genes that prevent cell growth, called tumor suppressor genes. Knowing which genes are affected in a tumor can help doctors to tailor the treatment for a particular cancer. Genetic information also predicts cancer risk for other family members.

RELATED TOPICS

See also
THE CELL CYCLE
page 48

MUTATIONS &
POLYMORPHISMS
page 68

DNA DAMAGE & REPAIR
page 70

3-SECOND BIOGRAPHIES

THEODOR HEINRICH BOVERI
1862–1915
German biologist who first proposed the cellular processes that cause cancer

ALFRED GEORGE KNUDSON
1922–2016
American physician who first hypothesized how accumulated mutations lead to cancer

HARALD ZUR HAUSEN
1936–
Nobel Prize-winning German virologist who discovered that HPV can cause cervical cancer

30-SECOND TEXT
Jonathan Weitzman

3-SECOND THRASH
Cancer is the most common human genetic disease, affecting young and old, rich and poor alike.

3-MINUTE THOUGHT
Although some cancer mutations are inherited, most cancers are caused by genetic changes that occur during an individual's lifetime. Infections may be responsible for as many as one-fifth of all cancers. Experts predict that more than 30 percent of cancers could be prevented, by reducing tobacco smoking, improving healthy living, and immunization against viral infections. Cancer genetics has taught us a lot about how normal cells divide and grow.

Understanding the genes involved is key to cancer treatment.

TECHNOLOGIES & EXPERIMENTAL APPROACHES

allelles Alternative variant forms of a gene that result from a mutational change in DNA sequence or expression of the gene. Alleles can be recessive, meaning they only have an effect when there are two copies, or dominant, where a single copy is enough to have an effect.

apoptosis Cell-suicide program found in multicellular organisms. It is a highly regulated process involving biochemical events that cause cell death. It is important for development; billions of cells can die every day in embryos and children. It is also used to remove damaged cells.

chromosome pairs Long strings of DNA that carry genes and genetic information. In eukaryotic cells (those with a discrete nucleus) the chromosomes are in the nucleus and composed of DNA, some RNA, and proteins. A prokaryotic cell (one without a discrete nucleus) has a single chromosome made entirely of DNA. Autosomes are chromosomes that are not sex chromosomes (X or Y). Autosomes exists in pairs, each carrying the same genes.

cystic fibrosis Genetic disease that principally affects the lungs (as well as some other tissues). Patients have difficulty breathing and frequent lung infections. The disease is inherited in an autosomal recessive manner. Parents can be carriers as they have only one copy of the CFTR gene that is mutated, whereas patients received two mutated copies, one from each parent.

DNA markers Gene or DNA sequence from a known location on a chromosome that is used to identify individuals or species.

DNA microarrays Miniature technology to measure the expression levels of many genes simultaneously or to study multiple regions of a genome. Specific pieces of DNA are used as probes and are spotted on a solid surface and then samples of DNA or RNA are tested to see where they stick (or hybridize). Microarrays are sometimes called "DNA chips."

DNA polymerase Enzyme that synthesizes DNA from nucleotides, the building blocks of DNA using a strand of DNA as a template. Cells use DNA polymerase to duplicate their genome before cell division. Researchers use DNA polymerase in the lab to copy pieces of DNA for cloning experiments.

DNA sequencing Technology to determine the precise order of nucleotides (A, G, C, or T) within a DNA molecule. The initial methods were slow and laborious, but modern methods are rapid and automated. DNA sequencing is now used extensively in medical diagnostics, biotechnology, and forensic science.

eukaryote Organism composed of cells with a distinct nucleus and cytoplasm. There are also living cells without a nucleus, such as bacteria, called "prokaryotes." Eukaryotes can be unicellular, such as yeast. Or they can be multicellular, such as humans.

feedback loops Auto-regulating system in which the output of a pathway is used to regulate the initial process, thereby creating a circuit or loop. Feedback loops can be negative or positive, either curtailing or reinforcing the signal.

genome The complete set of genetic material within an organism or a cell. Genomics is the study of an organism's genome, focusing on its evolution, function, and structure.

genotype and phenotype The genotype is the DNA sequence of a cell or an organism that determines a specific characteristic (a phenotype) of that cell or organism. The phenotype is the observable characteristics of a cell or an organism (such as shape, development, biochemical or physiological features, or particular behaviors).

nucleotides Building blocks used to make DNA or RNA. Strings of nucleotides are called nucleic acids. In DNA there are four nucleotides (referred to by the letters T, C, G, and A) and in RNA there are four ribonucleotides (U, C, G, and A). Nucleotides are also called bases. DNA bases can be paired: A pairs with T, and C pairs with G.

recombination frequency Measure of the genetic distance between two loci in order to create a genetic linkage map. Recombination frequency is the frequency of a single chromosomal crossover event between two genes during meiosis.

single-nucleotide polymorphisms (SNPs) Variation in a single nucleotide at a specific position in the genome, where each variation is present to some appreciable degree within a population. Many diseases result from SNPs, especially if the two variants affect protein structure and function.

MODEL ORGANISMS

the 30-second theory

All living organisms use DNA

as their genetic material, so we can learn a lot about genetics by studying almost any species. Studies in non-human species serve as models for understanding development and biological processes in other organisms. They are also useful for exploring the underlying mechanisms of physiology and disease in humans. Model organisms are selected because they are easy to maintain and breed in the laboratory. Some have particularly short life cycles, so several generations can be studied quickly. Researchers also choose model organisms with traits that are easily measured, such as body size and lifespan. There are now many model organisms in laboratories. One of the first was the bacterium *Escherichia coli* (*E. coli*), which was used to decipher the basic mechanisms of gene regulation. Single-cell organisms such as "baker's yeast" *Saccharomyces cerevisiae* helped scientists to understand genetics and cell biology. Human proteins that control the cell cycle can even replace those in yeast. The fruit fly *Drosophila melanogaster* has proved invaluable for studying developmental processes. And the roundworm *Caenorhabditis elegans* taught us how cells die by a conserved, programmed suicide process. Mice with specific gene mutations are studied as powerful models for human diseases.

3-SECOND THRASH
Non-human species are invaluable as tools in the laboratory for exploring gene functions.

3-MINUTE THOUGHT
Researchers use model organisms to perform genetic screens, searching for genes that affect specific phenotypes. The experiment can also be done the other way round, by creating genetically modified model organisms in the laboratory and looking at the outcomes on different traits. The latter is called "reverse genetics." The consequences of gene mutations and biological mechanisms are often conserved across species.

RELATED TOPICS
See also
GENOTYPE & PHENOTYPE
page 62

GENETICALLY MODIFIED ORGANISMS
page 146

3-SECOND BIOGRAPHIES
THOMAS HUNT MORGAN
1866–1945
American geneticist who used the fruit fly *D. elanogaster* to show that genes are carried on chromosomes

SYDNEY BRENNER
1927–
South African-born biologist who proposed the nematode worm *C. elegans* as a model for neuronal development

PAUL MAXIME NURSE
1949–
English geneticist who showed that the genes that control cell division are conserved from yeast to humans

30-SECOND TEXT
Jonathan Weitzman

Certain species of fruit fly, frog, and worm are all used as model organisms.

GENETIC FINGERPRINTING

the 30-second theory

3-SECOND THRASH
The analysis of a few genes with specific characteristics is sufficient to generate unique genetic fingerprints for paternity testing, crime scene analysis, and the identification of remains.

3-MINUTE THOUGHT
The British geneticist Alec Jeffreys first recognized the potential of genetic fingerprinting in the 1980s to prove a familial relationship in a case of disputed paternity and to identify the person responsible in rape and murder cases. Jeffreys' original approach was developed to include standardized and reproducible genetic analyses of many genes. Today it is used worldwide for all circumstances requiring individual genetic identification.

Our fingerprints are unique to each of us, and the same uniqueness is found in our DNA. Like detectives at a crime scene, geneticists use different types of polymorphic DNA sequences to capture the genetic fingerprints of people and animals. The DNA sequences used in genetic fingerprinting are carefully selected to ensure that each gene has many alleles and that the frequencies of all alleles are known in all populations. The genes selected each have one allele inherited from each parent. Genetic fingerprinting is used to identify paternity in disputed cases. Any fingerprinting allele that a child does not inherit from the mother must come from the father. This means when the father's genotype is tested, any non-maternal allele carried by the child must exist in the father's genotype. Genetic fingerprinting can identify human or animal remains, or biological material collected at a crime scene. Forensic scientists determine the genotypes for each fingerprinting gene and then calculate the probability that a person carries the particular set of genotypes for all genes tested by multiplying the frequencies of the genotypes. Often, the probabilities obtained are so small that effectively only one person in the world has a particular set of genetic fingerprinting genotypes.

RELATED TOPICS
See also
MENDEL'S LAWS
OF HEREDITY
page 16

MUTATIONS
& POLYMORPHISMS
page 68

3-SECOND BIOGRAPHIES
ALEC JEFFREYS
1950–
British geneticist who developed the first genetic fingerprinting methods and applied them to cases of disputed paternity and to the analysis of crime scene material

PETER NEUFELD
& BARRY SCHECK
1950– & 1949–
American lawyers who founded the Innocence Project, an organization dedicated to exonerating wrongly convicted people by applying DNA fingerprinting

30-SECOND TEXT
Mark Sanders

Genetic fingerprinting has many uses, including establishing a person's paternity.

GENETIC TESTING

the 30-second theory

Genetic tests look for mutations

in DNA or for abnormalities in blood proteins that indicate genetic diseases. Genetic testing is conducted at different ages and for different reasons depending on the particular case. It has been routine in hospital clinics since the 1970s. Prenatal genetic testing usually examines either DNA or chromosomes, looking for mutations. Chromosome analysis looks for extra or missing chromosomes or chromosome segments. Newborn genetic testing examines blood taken from babies in the first day of life, looking for signs of around 50 rare but treatable genetic diseases. Some genetic diseases in newborns can be treated by diet and medication to prevent disease symptoms or to reduce disease severity. Genetic testing in older patients can confirm a clinical suspicion of disease, identify which mutation a person carries, or identify persons who are heterozygous carriers of a gene mutation. Ideally, genetic testing can identify people at risk for carrying a disease-causing mutation before any disease symptoms appear. For example, genetic testing for mutations increasing the risk of certain cancers can change the way doctors treat patients. Today, some personalized genomic companies are proposing direct-to-consumer genetic tests.

RELATED TOPICS
See also
MUTATIONS
& POLYMORPHISMS
page 68

DOMINANT & RECESSIVE
GENETIC DISEASES
page 104

PERSONALIZED
GENOMICS & MEDICINE
page 140

3-SECOND BIOGRAPHIES
ROBERT GUTHRIE
1916–55
American microbiologist who developed the "Guthrie Test" to detect the treatable genetic disease phenylketonuria (PKU) in newborn babies

FRANCIS COLLINS
1950–
American former director of the Human Genome Project who made many contributions to our understanding of genetic diseases

30-SECOND TEXT
Mark Sanders

Testing in fetuses and newborn babies can reveal the presence of genetic diseases.

3-SECOND THRASH
Genetic testing seeks to identify gene mutations, abnormal blood proteins, or chromosomal changes that are associated with genetic diseases.

3-MINUTE THOUGHT
Genetic tests can detect abnormalities in proteins, chromosomes, or genes, but this information must be interpreted by experts and carefully explained to patients or families. Some genetic diseases detected in newborns can be treated immediately. In some cases, the detection of a mutation and clinical disease diagnosis might lead to additional testing of other family members to see if they carry the same mutation. The identification of a mutation linked to cancer risk offers the chance to closely monitor for disease development.

GENETIC MAPS

the 30-second theory

3-SECOND THRASH
A genetic map gives the order of genes and the distance between genes on a chromosome.

3-MINUTE THOUGHT
Genes were traditionally thought of as units of heredity that influence physical traits. But many different segments of DNA can be genes. For example, the human genome contains millions of locations at which variation of single base pairs occurs between different people. These so-called single nucleotide polymorphisms, or SNPs, are transmitted just like the genes that control physical traits. SNPs are genetic DNA variations that were extensively used to generate detailed genetic maps of chromosomes and they helped place the genes on these maps.

Maps offer useful tools to navigate landscapes. Genetic maps provide a guide to how the genes are laid out on a chromosome. Genes that are far apart from one another or located on separate chromosomes obey Mendel's law of independent assortment. But genes that are near one another on the same chromosome are genetically linked and do not separate independently. The alleles of linked genes on a chromosome tend to stay together during hereditary transmission. The alleles of linked genes are only separated when they are reshuffled during recombination between chromosome pairs. Geneticists determine the frequency with which alleles of linked genes are transmitted together and the frequency of recombination that separates them. In general, the higher the frequency of recombination between a pair of genes the more distant they are from one another. Lower recombination frequencies correspond to genes that are closer together. Geneticists use recombination frequencies to estimate the distance between genes and the order of genes along a chromosome. Like the cities and towns located along a road, the genes on a chromosome are mapped by their order and distance. Genetic maps played a critical role in the first genome sequencing projects.

RELATED TOPICS
See also
MENDEL'S LAWS
OF HEREDITY
page 16

THE HUMAN GENOME
PROJECT
page 30

GENETIC FINGERPRINTING
page 120

3-SECOND BIOGRAPHIES
THOMAS HUNT MORGAN
1866–1945
American geneticist who first hypothesized genetic linkage

ALFRED STURTEVANT
1891–1970
American geneticist who devised the first genetic map

THE INTERNATIONAL SNP
MAP WORKING GROUP
1998–2001
International consortium that mapped over 1.4 million SNPs in the human genome

30-SECOND TEXT
Mark Sanders

Like geographic maps, genetic maps show where key "landmarks" (the genes) are located.

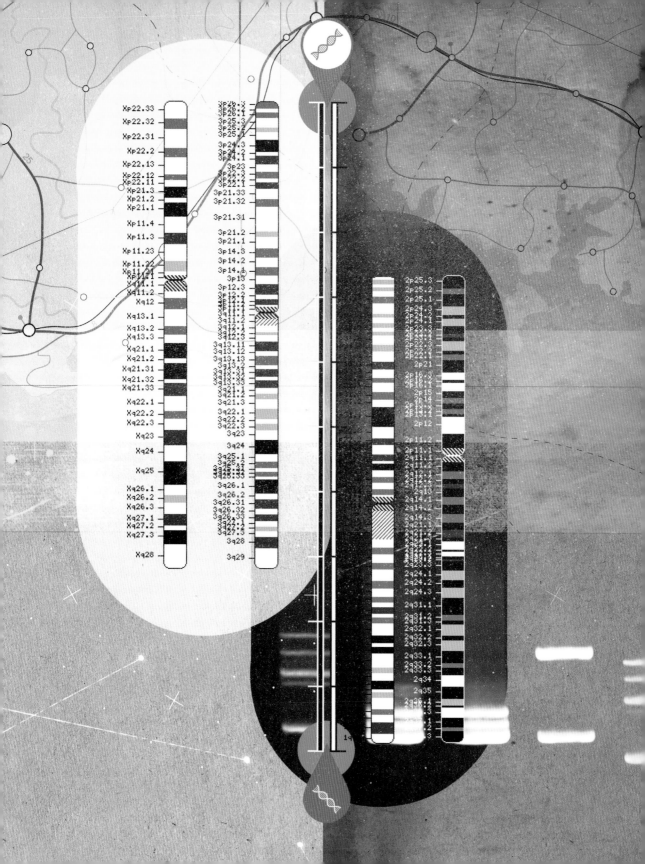

DNA SEQUENCING

the 30-second theory

Imagine a story that is one million pages long with 3,000 letters per page. That's the story of your DNA. The DNA molecule is a long string of letters called nucleotides. At each position there are one of four letters: A for adenine, T for thymine, G for guanine, and C for cytosine. A four-letter alphabet may seem simple, but consider that your DNA is three billion letters long. These letters make up the many different genes that control your traits. DNA sequencing has been called the most important tool in molecular biology. Sequencing technology can determine the order of nucleotides along the DNA. Thousands of researchers and clinicians sequence DNA to study how genes work and to understand how changes in the letters cause diseases, such as cancer and cystic fibrosis. DNA sequencing also provides information about the level of genetic variation in selected populations. In 2015, researchers sequenced the DNA of more than 2,500 people from around the world and compared the order of their letters. The results showed that, genetically speaking, humans are extremely similar to each other. If you pick two unrelated people at random, their DNA sequences differ by only 0.15 percent. In other words, our DNA sequences are 99.85 percent the same.

3-SECOND THRASH
DNA sequencing technology allows researchers to determine the order of the nucleotide letters in DNA.

3-MINUTE THOUGHT
If you compared the DNA sequences between two Japanese people or between a Japanese person and a Norwegian, the differences between both pairs would be around 0.15 percent. The value for the first pair (Japanese–Japanese) might be a little lower (say, 0.14 percent) and the other pair (Japanese–Norwegian) might be a little higher (say, 0.16 percent), but both pairs would be remarkably similar. Relatively little additional genetic variation is observed when comparing individuals from populations that are geographically separated.

RELATED TOPICS
See also
THE DOUBLE HELIX
page 22

THE HUMAN GENOME PROJECT
page 30

MUTATIONS & POLYMORPHISMS
page 68

3-SECOND BIOGRAPHIES
FREDERICK SANGER
1918–2013
British biochemist who invented one of the first methods for sequencing DNA, earning him the Nobel Prize in Chemistry in 1980. He also won it in 1958 for work on the structure of proteins

WALTER GILBERT
1932–
American biochemist who pioneered DNA sequencing techniques and promoted the Human Genome Project

30-SECOND TEXT
Robert Brooker

DNA sequencing determines the precise order of the nucelotides A, T, G, and C.

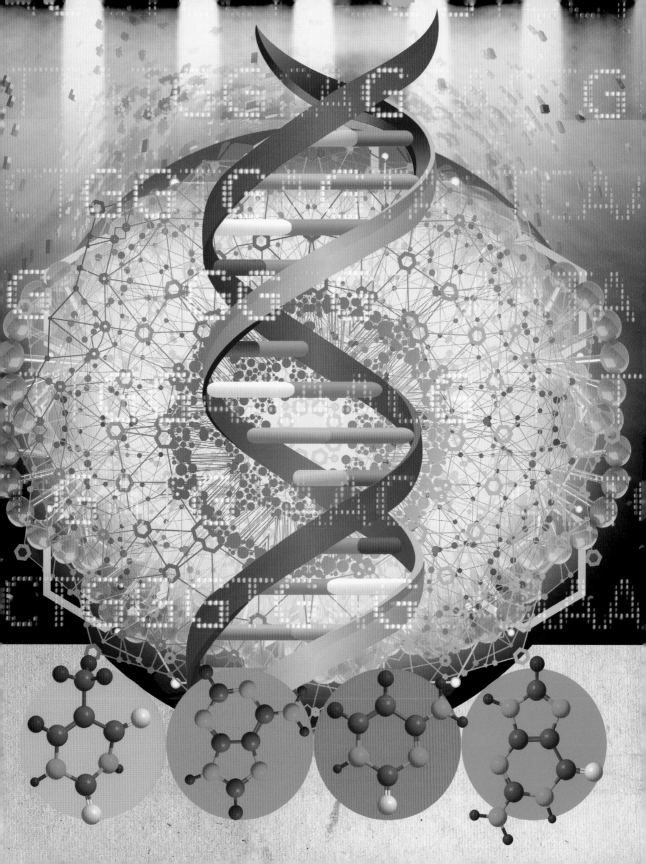

February 9, 1910
Born in Paris, France

1928
Begins his studies in
biology at the Sorbonne
(University of Paris)

1938
Marries archaeologist and
orientalist Odette Bruhl

1941
Receives doctorate from
the Sorbonne

1942–45
Enters the French
Resistance movement
during the Second World
War, eventually becoming
chief of the national staff

1945–76
Works at the Pasteur
Institute in Paris,
where he conducts
his famous studies
on gene regulation

1960
Becomes an honorary
foreign member of the
American Academy of
Arts and Sciences

1965
Awarded the Nobel Prize
in Physiology or Medicine

1970
Publishes his famous
book *Chance and
Necessity: Essay on
the Natural Philosophy
of Modern Biology*

1971
Becomes director of
the Pasteur Institute

May 31, 1976
Dies of leukemia
and is buried on the
French Riviera

JACQUES MONOD

Jacques Lucien Monod was born in Paris in 1910 to an American mother and a French father, Lucien Monod—a painter who was an intellectual inspiration to Jacques.

In 1928, Monod started his studies in biology at the Sorbonne, but soon came to realize that biology education there was lagging behind contemporary research. He indicated that he also learnt a great deal from others, a few years older than himself, outside of the university, that contributed to his true understanding of biology. He went on to obtain his science degree in 1931 and later began to work on bacterial growth, starting again at the Sorbonne in 1937 and receiving his doctorate degree in 1941. Monod had deep political convictions and during the Second World War he was active in the French Resistance, becoming chief of staff of operations for the Forces Françaises de l'Interieur and coordinating parachute drops ahead of the Allied landings.

After the war, he joined the staff at the Pasteur Institute in Paris, where he famously discovered how genes are regulated—that is, how they are turned "on" and "off" in response to environmental changes. Monod and his colleague François Jacob studied how genes in bacteria are regulated by lactose (a type of sugar) in their environment. They identified a key regulator, known as the "lac repressor," which is able to turn off lactose-utilizing genes when lactose is absent from the environment. For this work, Monod was awarded the Nobel Prize in Physiology or Medicine in 1965, sharing it with colleagues François Jacob and André Lwoff, who studied gene regulation in viruses.

Monod is also famous for proposing that a certain type of RNA acts as a genetic messenger to provide the information for protein synthesis from the DNA to the ribosome. He hypothesized that this RNA, which he called "messenger RNA" (mRNA), is transcribed from the nucleotide sequence within DNA and then directs the synthesis of particular polypeptides (protein chains). This proposal was a remarkable insight, considering that it was made before the isolation and characterization of mRNA molecules in the laboratory.

Monod was also an accomplished musician and a thoughtful writer. In 1970, he published his philosophical essay "Chance and Necessity" that discusses the process of evolution and the essential role of enzymatic feedback loops to explain complex biological systems. Monod stated his belief that the ultimate aim of science is to "clarify man's relationship to the universe."

In 1971 Monod was appointed director of the Pasteur Institute, where he worked until his death from leukemia in 1976. He is considered by many as one of the founding fathers of molecular biology.

Robert Brooker

POLYMERASE CHAIN REACTION (PCR)

the 30-second theory

3-SECOND THRASH
Polymerase chain reaction (PCR) is a laboratory technique that makes many copies of DNA in a defined region.

3-MINUTE THOUGHT
PCR works because each strand of DNA contains the sequence information to make a reverse copy. PCR is used in labs to learn how a gene works and to identify mutations that cause disease, as well as to clone genes from one species into another and create transgenic organisms for the pharmaceutical and biotech industry. PCR is even used at crime scenes to amplify tiny amounts of DNA from blood stains or hair roots.

PCR stands for "polymerase chain reaction," a laboratory technique that copies DNA in a test tube. PCR is used to clone genes and to make many copies of a particular DNA region. Researchers begin with a sample of DNA, such as chromosomal DNA from a human cell, and add short DNA sequences (called "primers") that bind on either side of the gene they want to copy. The test tube also contains nucleotides (the building blocks of DNA) and DNA polymerase (the enzyme that connects nucleotides to make long DNA polymers). The DNA polymerase enzyme is often isolated from bacterial species that live in hot springs and can function at high temperatures. PCR involves three steps: first, the chromosomal DNA is heated to separate the DNA into its two strands; next, the primers bind to each DNA strand as the temperature is lowered; third, the temperature is slightly raised, and DNA polymerase uses nucleotides to synthesize new DNA strands in the defined region, thereby doubling the amount of DNA fragment between the two primers. These three steps are repeated many times in a row, which is why the method is called a chain reaction. In just a few hours, PCR can increase the amount of DNA by a billionfold.

RELATED TOPICS
See also
GENETIC FINGERPRINTING
page 120

GENETIC TESTING
page 122

CLONING
page 148

3-SECOND BIOGRAPHIES
ARTHUR KORNBERG
1918–2007
American biochemist who won the Nobel Prize in Physiology or Medicine in 1959 for his discovery of DNA polymerase

KARY MULLIS
1944–
American biochemist who is largely credited for developing the technique of PCR, for which he won the Nobel Prize in Chemistry in 1993

30-SECOND TEXT
Robert Brooker

PCR works by heating and cooling DNA strands at specific temperatures to start a chain reaction.

GENOME-WIDE ASSOCIATION STUDIES (GWAS)

the 30-second theory

RELATED TOPICS

See also
GENETIC MAPS
page 124

PERSONALIZED GENOMICS
& MEDICINE
page 140

3-SECOND THRASH

Genome-wide association studies take advantage of the close association of specific genetic markers to DNA variations that alter our characteristics or phenotypes.

3-MINUTE THOUGHT

GWAS measure the association of hundreds of thousands of genomic DNA markers to specific characteristics such as susceptibility to a disease, height, weight, and so on. Often the associated DNA changes only explain a small percentage of the phenotypic variation— for example, 0.4 in (1 cm) of human height or 2.5 percent of susceptibility to develop a disease. The accumulation of small effects can have a significant impact on the phenotype. Such studies are applicable to plants and animals.

Any human genome contains thousands of variations in the sequence of DNA compared to anyone else's genome. This makes every individual unique. However, there are segments of DNA, with exactly the same sequence, that are shared by groups of people. Most differences in the DNA sequence do not modify characteristics such as height, weight, and so on. But some DNA changes do have an impact on traits and these can be embedded within DNA phrases that we are able to read. Genome-wide association studies (GWAS) use the differences between individuals to map the genetic causes of given traits. Using DNA microarray technologies, researchers measure the presence of hundreds of thousands of DNA variants in the genome. Many of these are changes in just one letter of DNA at a specific position. These are called single nucleotide polymorphisms (SNPs). SNPs are used as DNA markers because they can be next to a gene responsible for a particular trait. For example, comparing individuals with different heights, we might observe that shorter people have an A at a given SNP position, while taller people have a G. If genetic analysis shows that an individual has a G, then he or she is more likely to be taller. GWAS apply this to the whole genome to find variants associated with traits and diseases.

3-SECOND BIOGRAPHIES
DAVID BOTSTEIN
1942–
American biologist who proposed a method to construct genetic maps that paved the way for the association studies

ERIC LANDER
1957–
American geneticist who recognized, with David Botstein, the potential of DNA markers to study complex human traits and diseases

30-SECOND TEXT
Reiner Veitia

At a given DNA position there are variations in sequence between individuals. These differences might be related to a phenotype, such as height.

THERAPEUTIC PROMISE

THERAPEUTIC PROMISE
GLOSSARY

computational modeling Use of a computer to simulate the behavior of a biological system. Computer models can be useful to understand how systems are constructed and to test what happens when they are perturbed. The use of computational approaches in biology is called "bioinformatics."

CRISPR-Cas9 Latest technology for precise genome editing. The CRISPR ("clustered regularly interspaced short palindromic repeats") and Cas9 system was discovered in bacteria, where it acts as a primitive immune system to protect against invading genetic material from viruses. Its ability to recognize precise DNA sequences and cut them had been exploited to engineer a powerful tool for cut-and-paste techniques on eukaryote genomes.

expressed sequence tags (EST) Short subsequence of a cloned cDNA that can be used to identify gene transcripts for quantification and for gene discovery. They are relatively short fragments that represent bits of expressed genes.

germ line Cell that gives rise to the gametes for sexual reproduction. Germ cells undergo meiosis, followed by cellular differentiation to produce mature gametes, either eggs or sperm. Gametes contain the genetic information that will be transmitted to the next generation.

induced pluripotent stem cells (iPSC) Stem cells that were generated from normal adult cells by a process of reprogramming. iPSC can be differentiated into different cell types.

lentiviral vectors Modified viruses that are used to deliver genes for gene therapy. They are RNA viruses (for example, HIV) that can be engineered to carry genes that are delivered when the virus infects the patient cells.

metagenomics Study of genetic material obtained from environmental samples. DNA sequence analysis reveals the hidden diversity of microscopic life and the microbial world. Metagenomics has expanded rapidly due to the falling price of DNA sequencing technologies.

nucleases Enzymes that cut DNA. Researchers have engineered these natural enzymes so that they can target specific DNA sequences for genome editing. For example, Zinc-finger nucleases (ZFNs) use a special protein domain that recognizes precise DNA sequences. Researchers use ZFNs, together with TALEN and CRISPR-Cas9 technologies, to cut-and-paste DNA sequences and edit genomes.

oncogenicity The capacity to induce tumors. Genes that induce cancer are called oncogenes. Genes that prevent tumor formation are called tumor suppressor genes.

oocyte Female gamete (egg) or germ cell involved in reproduction. It is produced in the ovary during female gametogenesis. Cloning experiments use "enucleated eggs," which are oocytes in which the nucleus has been removed.

pluripotent Capacity of a stem cell to give rise to several different cell types. Pluripotent cells can generate all of the cell types that make up the body. Embryonic stem cells are considered to be pluripotent.

somatic cells Biological cells that form the main body of an organism. There are hundreds of different types of somatic cell types in the human body that make up the organs and tissues. Somatic cells are not transmitted to the next generation and are distinct from germ cells and gametes.

stem cells Undifferentiated cells that can differentiate to generate more specialized cell types. Embryonic stem cells can generate all the different cells in the embryo (they are pluripotent), whereas adult stem cells can normally only generate cells for specific tissues.

TALEN Enzymes that cut specific sequences of DNA. Their full name is "transcription-activator-like effector nucleases" because they are made by fusing transcription proteins to nucleases. They can be engineered to cut any desired DNA sequence and have become a powerful tool for genome editing.

transgenic organism Animal or plant generated by introducing a foreign gene (a transgene) or DNA. The transgene can change the characteristics (phenotype) of the organism. Sometimes referred to as genetically modified organisms and the source of much public debate about safety issues.

virus Small infectious agent that can replicate only inside living cells. Viruses can infect all types of life, including animals and plants and bacteria. The study of viruses is called virology. Viral particles, called virions, contain genetic material (DNA or RNA) and an outer protective coat called the capsid. Most viruses are so small that they cannot be seen with a normal light microscope.

GENE THERAPY

the 30-second theory

3-SECOND THRASH
Gene therapy inserts genetic material into cells to confer new characteristics, to correct genetic diseases, or to strengthen defenses against cancer.

3-MINUTE THOUGHT
Recently, researchers developed innovative tools (called engineered nucleases) that can cut the genome at very specific positions, promising new possibilities for precise intervention into the genome. These molecular machines make it easy to disrupt a gene or add DNA in a specific place. Such strategies could correct a genetic disorder by replacing the mutated DNA with a normal sequence, while leaving the gene in its physiological context.

When researchers realized that some diseases are caused by mutations in single genes, they proposed that gene therapy could be used to correct the problem gene with a normal copy. It might even be possible to add genes to change the properties of a given cell. In most cases gene therapy uses a vector (delivery agent) to transfer the therapeutic gene into the target cell. Viruses are the most effective vectors, because they persist and can often also integrate into the host genome. There are still obstacles to efficient and safe gene therapy, including the difficulty of making sure the genetic material is safely put to use and that the genes do not provoke an immune response in the body or lead to the formation of tumors. So far, gene therapy has successfully treated inherited conditions of the hematopoietic system (the organs that make blood), including severe immunodeficiencies and leukodystrophies (genetic diseases that affect the brain, spinal cord and the peripheral nerves). And progress was reported in gene therapy of hemophilia B and inherited retinal dystrophies (which causes progressive blindness). Gene therapy of the future will benefit from technological advances in vector design and production. Gene therapy is making progress to treat more complex diseases, such as cancer.

RELATED TOPICS
See also
GENES & IMMUNODEFICIENCY
page 108

PERSONALIZED GENOMICS
& MEDICINE
page 140

GENOME EDITING
page 152

3-SECOND BIOGRAPHY
LUIGI NALDINI
1959–
Italian doctor who developed lentiviral vectors to be used in gene transfer

30-SECOND TEXT
Alain Fischer

Viruses invade cells and can integrate their DNA into the host cell's genome. This makes them ideal as a vector (or delivery agent) to use in DNA therapy.

PERSONALIZED GENOMICS & MEDICINE

the 30-second theory

The Human Genome Project

fueled the emergence and development of DNA sequencing technologies, which can efficiently decipher the sequence of any genome. The cost of sequencing a human genome has dropped dramatically, from billions of dollars two decades ago to just $1,000 today. This progress makes it possible to access the sequence of our own genomes. Whether we like it or not, we live in an era of personal genomics and the quantified self. A genome sequence gives access to our past, but also, to some extent, to our future. Our DNA carries genetic variations from our ancestors and tells us something about their origins. Other variations are not neutral and can have consequences for our health. "Reading" the genome can provide clues about where we come from, but also about risks of developing a disease, depending on the influence of the environment and other gene variants. A practical application of personal genomics is "personalized medicine." Until recently, most drugs were prescribed on the assumption that they would be effective for everyone. But by analyzing the genome sequence of an individual we can select more appropriate therapies and tailor dosages to prevent adverse side-effects.

3-SECOND THRASH
Personal genomics and personalized medicine promise to improve individual treatment because we now have access to the human genome sequence at a reasonable cost.

3-MINUTE THOUGHT
Access to individual genome sequences at an affordable cost raises numerous ethical issues. Analysis of genomic sequences is closely regulated by law in many countries to avoid any potential genetic discrimination. Having access to our own genome sequence can also be stressful. In a few clear-cut cases, a specific DNA variation is likely to impact our health. But, in most cases DNA variations only imply a potential risk.

RELATED TOPICS
See also
THE HUMAN
GENOME PROJECT
page 30

GENETIC TESTING
page 122

GENETIC MAPS
page 124

3-SECOND BIOGRAPHIES
J. CRAIG VENTER
1946–
American biologist who played an important role in the race to sequence the human genome

FRANCIS COLLINS
1950–
American geneticist and leader of the Human Genome Project

30-SECOND TEXT
Reiner Veitia

Understanding an individual's genome can allow doctors to tailor specific treatments for a range of diseases, including cancer.

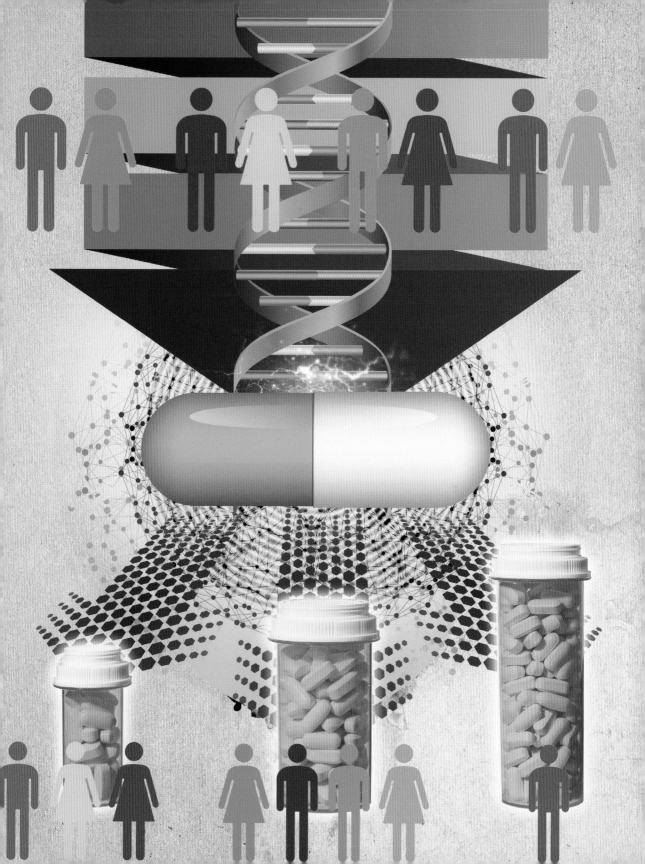

SYNTHETIC BIOLOGY

the 30-second theory

Synthetic biology is a relatively new field and scientists have many different definitions of what it is. One underlying theme is the application of engineering principles to the components of a cell, to elicit a particular action in response to an input. Advances in biotechnology and the computational modeling of biological processes enable us to manipulate existing genetic or biochemical pathways or to create artificial ones. These engineering techniques hold for molecules, cells, tissues, and organisms. For example, designing an enzyme that is capable of cutting DNA at a specific sequence could be considered synthetic biology. The replacement of a DNA or protein component in a living organism by a non-natural component is also synthetic biology. Bacteria that shine light in response to a chemical in the culture medium or bacteria able to kill tumor cells are also products of synthetic biology. The common underlying theme is that the universality of the genetic code allows scientists to engineer new DNA sequences that will confer new properties to the recipient cells. This emerging discipline raises hope—because it is now possible, at least in principle, to create living organisms for many new applications.

3-SECOND THRASH
Synthetic biology involves the rational design of biological components and systems, based on knowledge of the biochemistry and functions of natural organisms.

3-MINUTE THOUGHT
The advent of synthetic biology raises ethical questions. There are concerns about potential threats to the health of living organisms or the environment if an engineered molecule or organism were to escape from a research laboratory. There are also concerns about the fairness and the (in)appropriateness of owning patents on a living organism and its components.

RELATED TOPIC
See also
CRACKING THE GENETIC CODE
page 24

GENETICALLY
MODIFIED ORGANISMS
page 146

3-SECOND BIOGRAPHIES
STÉPHANE LEDUC
1853–1939
French biologist and chemist, first to use the term "synthetic biology," in 1910

GEORGE CHURCH
1954–
American geneticist who has played important roles in the fields of personal genomics and synthetic biology

30-SECOND TEXT
Reiner Veitia

Synthetic biology is being used to create artificial nucleic acids, which could help scientists to answer questions about the origins of life itself.

October 14, 1946
Born in Salt Lake City, Utah

1972
Receives his degree in biochemistry at the University of California, San Diego (UCSD)

1975
Receives PhD in physiology and pharmacology from UCSD

1976–84
Member of the faculty at State University of New York, Buffalo

1984–92
Section chief at the National Institute of Neurological Disorders and Stroke at the National Institutes of Health (NIH) in Bethesda, Maryland

1992
Founds the Institute for Genomic Research

1995
Sequences the first bacterial genome

1998
Founds Celera Genomics, Inc.

June 26, 2000
Jointly announces the mapping of the human genome, with Francis Collins of the NIH

2001
Publishes first draft of the human genome sequence

2002
Becomes president of the J. Craig Venter Institute, and the CEO of Human Longevity, Inc.

2010
Introduces a synthetic genome into a bacterial cell

J. CRAIG VENTER

The geneticist J. Craig Venter

is a pioneer of DNA sequencing technologies who played a major role in the first attempt at sequencing the entire human genome. He has been challenging and provoking the genetics community for more than two decades, questioning established approaches, and driving technology innovation. This earned him the title "Gene Maverick."

Born in 1946 in Salt Lake City, Utah, Venter studied at the University of California, San Diego, then in 1976 was appointed assistant professor at the State University of New York at Buffalo. There his research focused primarily on receptors involved in cell signaling. From 1984–92, as section chief at the National Institute of Neurological Disorders and Stroke at the National Institutes of Health, he developed gene tagging methods.

Leaving the NIH in 1992, Venter became founder and chairman of the board of The Institute for Genomic Research (TIGR), a not-for-profit genomics research institution. In 1998, Venter joined Applera Corporation and became president and chief scientific officer of the newly founded Celera Genomics, which focuses on genetic sequencing and related medical and biological information. Celera's corporate motto was "Speed Matters." These events led to a sequencing race between Celera and the National Institutes of Health-led Human Genome Project coordinated by Francis Collins. In 2001, both the private and public initiatives simultaneously published the first draft sequence of the human genome.

Venter's other key accomplishments include the sequencing of the first bacterial genome, *Haemophilus influenza*, and the first living organism in 1995. In his early days at the NIH, Venter developed novel methods for tagging genes by focusing on expressed genes that represent a small percentage of the human genome. These sequences, called ESTs (expressed sequence tags), led to massive gene discovery and raised legal questions about whether these new genes could be patented.

Venter and his collaborators have also studied environmental DNA samples, creating a new field called "metagenomics." In 2010, his research team made a synthetic DNA molecule and transferred it into a bacterial cell, thereby creating the first self-replicating bacterial cell constructed entirely with synthetic DNA.

Venter was listed in 2007 and 2008 in *Time* magazine's top 100 most influential people in the world, and in 2010 *New Statesman* magazine listed Venter as among the world's 50 most influential figures. He is a member of prestigious scientific organizations including the National Academy of Sciences, the American Academy of Arts and Sciences, and the American Society for Microbiology.

Robert Brooker

GENETICALLY MODIFIED ORGANISMS

the 30-second theory

RELATED TOPICS
See also
GENE THERAPY
page 138

CLONING
page 148

GENOME EDITING
page 152

3-SECOND THRASH
A GMO is any organism whose genetic material has been altered using genetic engineering techniques.

3-MINUTE THOUGHT
Although GMOs are rigorously tested for their safety, there has been considerable public debate about them. Some advantages of GMOs include healthier crops that require less pesticides, plants with improved nutritional value, or the production of expensive drugs like human insulin in bacteria. Some disadvantages might include allergies and the threat of genes spreading to other organisms.

Imagine a mouse that gives off

an eerie green glow like a jellyfish or a bacterium that makes human insulin. Although it sounds like science fiction, researchers have actually learned how to exchange genetic information to create these examples of genetically modified organisms (GMOs). Gene cloning and gene engineering techniques enable the introduction of genetic material from one species into another to create genetically modified bacteria, animals, or plants. When a GMO has genetic material from a different donor species, it is called a transgenic organism. The mouse that gives off an eerie green glow is a transgenic organism: researchers cloned a gene encoding a green fluorescent protein (GFP) normally expressed only in jellyfish and created a transgenic mouse that now expresses GFP and shines green like a jellyfish. Today many economically important examples of GMOs are in the field of agriculture. These include Bt-corn and Bt-cotton, which carry a gene from the bacterium *Bacillus thuringiensis*. This gene encodes a toxin that kills corn borers and other insects. These Bt varieties of plants produce the toxins themselves and are resistant to many types of caterpillars and beetles.

3-SECOND BIOGRAPHIES
HERBERT BOYER
& STANLEY COHEN
1936– & 1935–
American biotechnologist and American geneticist who in 1973 made the first GMO by removing an antibiotic resistance gene from one bacterium and inserting it into another bacterium, enabling that bacterium to survive in the presence of the antibiotic

RUDOLF JAENISCH
1942–
American geneticist who created transgenic mammals by introducing foreign DNA into early mouse embryos

30-SECOND TEXT
Robert Brooker

GMOs can have many benefits, despite being a source of controversy.

CLONING

the 30-second theory

The word "clone" means to make many identical copies of something. In genetics, the term "gene cloning" refers to making a molecular copy of a gene. Genes can be cloned using a laboratory technique called polymerase chain reaction (PCR), in which the copying is done by an enzyme called DNA polymerase. An alternative way is to insert a gene into a plasmid (a circular DNA molecule that can copy itself independently of a cell's chromosomal DNA) and then put the plasmid into a living host cell, such as a bacterium or yeast cell. When the host cells divide and increase in number, many copies of the cloned gene are made, too. But cloning can also be at the level of whole cells or even whole organisms. Identical twins are clones that develop from the same fertilized egg. This cloning happens by accident, when a fertilized egg divides into two cells that separate from one other, each developing into a person with the same genetic material as the other. Researchers developed ways to clone whole mammals in the laboratory. They removed the DNA from oocytes and then fused the oocyte with a cell from the individual to be cloned. This process is called reproductive cloning and the first cloned mammal was a sheep named Dolly.

RELATED TOPICS

See also
WHAT IS A GENE?
page 56

TWINS
page 92

POLYMERASE CHAIN
REACTION (PCR)
page 130

3-SECOND BIOGRAPHIES
JOHN GURDON
1933–
English biologist who pioneered transplanting the nucleus from a tadpole cell into an enucleated frog egg. His famous experiments in the 1960s earned him the title "the Godfather of cloning" and a 2012 Nobel Prize

IAN WILMUT
1944–
English embryologist who with his colleagues at the University of Edinburgh produced clones of sheep using DNA from somatic cells

30-SECOND TEXT
Robert Brooker

Since Dolly, other mammals have been cloned, including pigs, horses, and deer.

3-SECOND THRASH
Cloning makes many copies of something and can be performed at the level of a gene, a single cell, or a whole organism.

3-MINUTE THOUGHT
Researchers have created clones of many mammalian species, including sheep, cows, mice, goats, pigs, and cats. In 2002, the first pet (a cat) was cloned. It was named CC (after the chemical symbol for carbon) and nicknamed "CopyCat." The cloning of mammals has many practical applications, including maintaining agriculturally valuable livestock and endangered species. However, a complete ban on human reproductive cloning was issued in many countries and research in this field is highly regulated.

STEM CELLS & REPROGRAMMING

the 30-second theory

3-SECOND THRASH

Stem cell research hopes to recreate specialized organs from any cell in the body and regenerate new tissues to restore functions affected by the aging process or disease.

3-MINUTE THOUGHT

One of the most important conclusions from stem cell experiments was that the genome is present and intact, in almost every somatic cell, not just in the germ line that normally transmits it to the next generation. However, what has really caught public attention is the idea that induced pluripotent stem cells and adult stem cells are potential elixirs of life that we can use to replace tissues worn out by old age or destroyed by disease.

All organisms have specialized stem cells that have the capacity to make many different cell types. Stem cells replenish organs when cells die or need to be replaced. For example, most of the cells in our intestines are lost and replaced every few days. But can any cell in our bodies change itself into another cell type or are cells programmed for just one cell type? Researchers were surprised to discover that almost all the cells in our bodies have the capacity to change into all other cells and have a spectacular capacity to be reprogrammed. Studies in the 1950s and 60s demonstrated that the nucleus of a cell could be reprogrammed by transferring it into an unfertilized egg that has had its nucleus removed. Indeed, these "cloned" eggs can even develop into embryos and sometimes to a fully grown adult. In 2006 scientists discovered the special conditions for cell reprogramming. They identified a cocktail of just four proteins that, when introduced into a cell, could generate "induce pluripotent stem" (iPS) cells. These stem cells have tremendous potential because they can be differentiated in a dish to create many types of cells and tissues. These iPS cells are used to study how development works, to model human disease and to produce cells and organs for regenerative medicine and tissue therapy.

RELATED TOPICS

See also
DEVELOPMENTAL GENETICS
page 100

CLONING
page 148

3-SECOND BIOGRAPHIES

JOHN GURDON
1933–
British developmental biologist who demonstrated that the nuclei of differentiated intestinal cells can generate into all of the cell types when reintroduced into an enucleated egg

SHINYA YAMANAKA
1962–
Nobel Prize-winning Japanese stem cell researcher who created "induced pluripotent stem (iPS) cells" after introducing four reprogramming factors into mouse fibroblasts

30-SECOND TEXT
Edith Heard

Stem cell research may lead to promising new treatments for diseases and major injuries.

GENOME EDITING

the 30-second theory

3-SECOND THRASH

The revolutionary technology of genome editing engineers the genome by targeting modifying enzymes to specific DNA sequences.

3-MINUTE THOUGHT

Researchers can now cut and paste genetic material with unprecedented precision. Enzymes targeted to specific DNA sequences in the genome can be used as "molecular scissors" to generate site-specific breaks. These breaks allow insertion or deletion of genomic sequences that can inactivate genes. Alternatively, new DNA can be provided to replace the endogenous gene with a targeted modification.

Using engineered enzymes to edit genomes is a new approach that promises to transform genetic studies and treatment of genetic diseases. It uses molecular tools to modify the genome at targeted points. This is achieved by creating site-specific enzymes called "nucleases" that establish DNA breaks at defined sequences in pieces of DNA: these include targeted zinc finger nucleases (ZFNs) and transcription-activator-like effector nucleases (TALEN). These approaches link a non-specific DNA cutting enzyme to proteins that recognize specific DNA sequences. An alternative technology is based on the microbial CRISPR-Cas9 system that harnesses RNA-programmed targeting of a nuclease to a specific DNA sequence. All these approaches introduce a specific double-strand break in DNA at a defined point in the genome. Once the genome is broken, repair enzymes can disrupt or replace DNA sequences at or near the point where the cut was made. The ability to modify the DNA sequence of a single cell or even whole organism in a targeted fashion allows studies that assess the impact of the change on the phenotype. Targeted nucleases also facilitate gene therapy for inherited disorders, the goal being to replace defective genes with normal alleles at the same natural location to correct the genetic mutation.

RELATED TOPICS

See also
GENE THERAPY
page 138

SYNTHETIC BIOLOGY
page 142

3-SECOND BIOGRAPHIES

EMMANUELLE CHARPENTIER
& JENNIFER DOUDNA
1968– & 1964–
French microbiologist and American chemist who in 2012 adapted the CRISPR system to take advantage of a synthetic guide RNA that directs the Cas9 enzyme

FENG ZHANG
1982–
Chinese-born biomedical scientist who in 2013 harnessed the CRISPR-Cas9 system for genome editing in eukaryotic cells

30-SECOND TEXT

Matthew Weitzman

Using engineered enzymes to edit genomes is an exciting new approach that could lead to major medical advances.

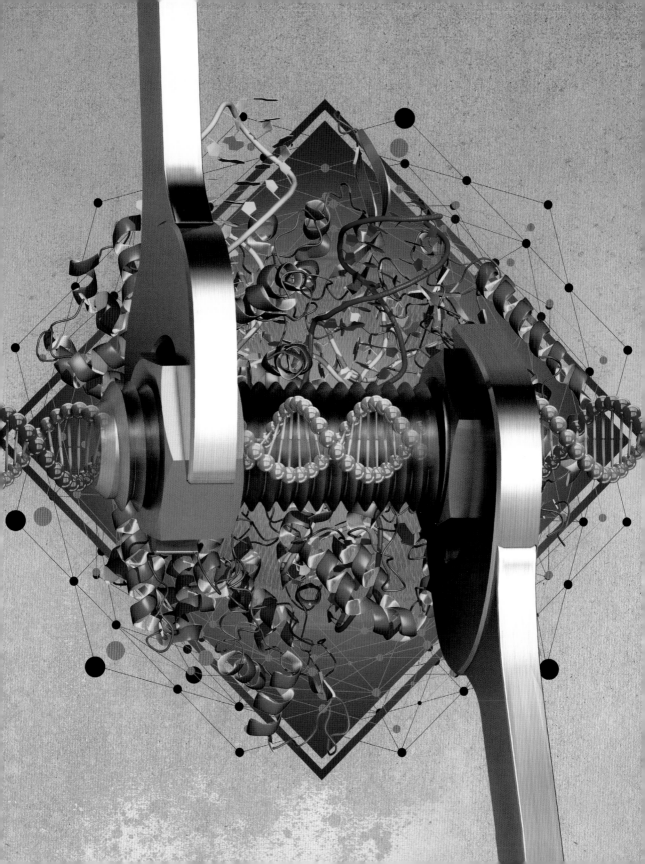

NOTES ON CONTRIBUTORS

EDITORS

Jonathan B. Weitzman is a professor of Genetics at the Université Paris Diderot and founding director of the Center for Epigenetics and Cell Fate. Jonathan teaches classes in genetics, epigenetics, and stem cell biology to students of all ages and he directs the European Masters' in Genetics program. His research focuses on understanding gene regulatory networks and epigenetic contributions to disease.

Matthew D. Weitzman is a professor at the University of Pennsylvania Perelman School of Medicine, and runs a lab in the Children's Hospital of Philadelphia. Matthew has a background in virology and molecular biology, and he studies the intersection between virus infection and genome integrity. He has lectured around the world and organized numerous scientific meetings on viruses, genome integrity, and gene therapy.

Jonathan and Matthew are identical twins.

FOREWORD

Rodney Rothstein is a professor of Genetics & Development and Systems Biology at Columbia University Medical Center. He is well known for his studies on DNA double-strand break repair and methods to edit genomes. His honors include the Genetics Society of America 2009 Novitski Prize, Doctor *honoris causa* in Medicine, Umeå, Sweden, and election to the American Academy of Arts and Sciences and the National Academy of Sciences.

CONTRIBUTORS

Thomas Bourgeron is a professor at the Université Paris Diderot in Paris. He leads a research group at the Institut Pasteur which gathers psychiatrists, neurobiologists, and geneticists together to study the biology of the social brain. One of his main findings has been the identification of a synaptic pathway associated with autism.

Robert J. Brooker received his PhD in genetics from Yale University in 1983. At Harvard, he studied lactose permease, the product of the *lacY* gene of the *lac* operon. He continued working on transporters at the University of Minnesota, where he is a professor in the Department of Genetics, Cell Biology, and Development. He is the author of several undergraduate genetics textbooks, including *Genetics: Analysis & Principles*, 6e, and *Biology*, 4e, published by McGraw-Hill Education.

Virginie Courtier-Orgogozo is a biology researcher at the Institut Jacques Monod in Paris. Her team studies the mutations responsible for differences between species in order to better understand our evolution, past, and future. She received the CNRS Bronze Medal and was elected "Young Woman Scientist" of the year, 2014, in France.

Alain Fischer is Professor at Collège de France, Paris, founding director of the Imagine Institute. He is a specialist of genetics and immunology notably of primary immunodeficiency diseases and gene therapy.

Edith Heard is a British geneticist, working on X-chromosome inactivation, with a long-standing interest in epigenetics, nuclear organization, and chromosome structure. She is director of the Genetics and Developmental Biology Unit at the Institut Curie in Paris and Professor of Epigenetics and Cellular Memory at the Collège de France. She is also a Fellow of the Royal Society.

Mark F. Sanders has been a member of the faculty in Molecular and Cellular Biology at the University of California, Davis since 1985, where he focuses on teaching genetics. He has also taught genetics at the University of Cambridge, England, and the University of Vienna. Mark is the lead author of *Genetic Analysis: An Integrated Approach* published by Pearson.

Reiner A. Veitia is a professor of Genetics at Université Paris Diderot. His research focus is primarily the genetics of female infertility and ovarian malignancies. He has also actively explored the molecular and theoretical bases of genetic dominance. He is the current Editor-in-Chief of the journal *Clinical Genetics* and a member of non-governmental association Academia Europaea.

RESOURCES

BOOKS

A Life Decoded: My Genome: My Life
J. Craig Venter
(Penguin Group, New York, 2008)

*A Monk and Two Peas: The Story of Gregor
Mendel and the Discovery of Genetics*
Robin Marantz Henig
(Phoenix, Orion Books, London, 2001)

*Chance and Necessity: An Essay on the
National Philosophy of Modern Biology*
Jacques Monod
(Fontana, HarperCollins, New York, 1974)

*Creation: The Origin of Life /
The Future of Life*
Adam Rutherford
(Penguin Group, New York, 2014)

*Epigenetics: How Environment
Shapes Our Genes*
Richard C. Francis
(W. W. Norton & Company, New York, 2012)

*Francis Crick: Discoverer
of the Genetic Code*
Matt Ridley
(HarperCollins, New York, 2006)

Genetics: Analysis and Principles
Robert R. Brooker
(McGraw-Hill Education, New York;
6th edn, 2016)

Genetic Analysis: An Integrated Approach
Mark F. Sanders and John L. Bowman
(Pearson, New York; 2nd edn, 2015)

*Here Is a Human Being:
At the Dawn of Personal Genomics*
Misha Angrist
(Perennial, HarperCollins, New York, 2011)

*Nature via Nurture: Genes, Experience
and What Makes Us Human*
Matt Ridley
(Perennial, HarperCollins, New York;
new edn, 2004)

On the Origin of Species
Charles Darwin
(Oxford University Press, New York; rev. edn, 2008)

*Redesigning Humans: Choosing
Our Genes, Changing Our Future*
Gregory Stock
(Houghton Mifflin (Harcourt), Chicago, IL, 2003)

Rosalind Franklin: The Dark Lady of DNA
Brenda Maddox
(HarperCollins, New York; new edn, 2003)

*The Double Helix: A Personal Account of
the Discovery of the Structure of DNA*
James Watson
(Penguin Books, New York; 2nd rev. edn, 1999)

*The Eighth Day of Creation: The Makers
of the Revolution in Biology*
Horace Freeland Judson
(Cold Spring Harbor Press, Cold Spring
Harbor, NY, 1979)

The Epigenetics Revolution
Nessa Carey
(Icon Books, London, 2012)

The Gene: An Intimate History
Siddhartha Mukherjee
(Scribner Book Company (Simon & Schuster,
New York), 2016)

*The Language of Life: DNA and the
Revolution in Personalized Medicine*
Francis S. Collins
(HarperCollins, New York, 2010)

*The Panda's Thumb: More
Reflections in Natural History*
Stephen Jay Gould
(W. W. Norton & Company, New York, 1980)

The Selfish Gene
Richard Dawkins
(Oxford University Press, New York, 1976)

*The Triple Helix: Gene, Organism,
and Environment*
Richard Lewontin
(Harvard University Press, Cambridge, MA, 2002)

WEBSITES

www.geneed.nlm.nih.gov
A free website for students, educators, and
interested citizens that offers up-to-date
information on genetics and biotechnology.

www.dnaftb.org
A primer on the 75 experiments that have
defined modern genetics, complete with
animations, interviews, and more.

www.learn.genetics.utah.edu
A free resource with a wide range of
explanations, activities, and experiments aimed
at anyone with an interest in genetics.

www.genome.gov
The website of the National Human Genome
Research Institute which participated in the
Human Genome Project.

www.ensembl.org/index.html
A genome browser for vertebrate genomes that
supports research in comparative genomics, and
provides tools for studying genome evolution,
variation, and regulation.

www.ncbi.nlm.nih.gov/omim
OMIM is the Online Mendelian Inheritance in
Man, providing an online catalog of human
genes and genetic disorders.

INDEX

ACKNOWLEDGMENTS

Jonathan Weitzman and Matthew Weitzman dedicate this book to Claire-Cipora and Sharon, with whom they juggled their genes.

PICTURE CREDITS

The publisher would like to thank the following individuals and organizations for their kind permission to reproduce the images in this book. Every effort has been made to acknowledge the pictures; however, we apologize if there are any unintentional omissions.

Alamy/evan Hurd: 144.

Getty/Bettmann / Contributor: 44, 66; Universal History Archive / Contributor: 26.

Science Photo Library/AMERICAN PHILOSOPHICAL SOCIETY: 90; HENNING DALHOFF: 39; GUNILLA ELAM: 87C; MARTIN KRZYWINSKI: Cover; US NATIONAL LIBRARY OF MEDICINE: 128.

Shutterstock/3drenderings: 103; 895Studio: 63T(BG); AbstractUniverse: 85BL, 85B; Aedka Studio: 81CL; Ahturner: 123CL; Alexilusmedical: 101B, 113B, 151CR; Alila Medical Media: 31C(BG); Anteromite: 2C, 121C; Aperture75: 79TC(BG); art_of_sun: 71C; Artos: 141T; Astronoman: 11C, 153C; ber1a: 71BG; Pedro Bernardo: 119BL; Bildagentur Zoonar GmbH: 71BL; BlueRingMedia: 23C; Evgenii Bobrov: 43T; gualtiero boffi: 103C; Olga Bogatyrenko: 57BC; Yevgeniya Bondarenko: 57TR; BortN66: 71BL(BG); Amanda Carden: 71TL(BG), 79TC; Catalinr: 127BG; Pavel Chagochkin: 99L&R, 111TC; Efstathios Chatzistathis: 147BR; Cherezoff: 143B(BG); Chromatos: 29T(BG); Cico: 11C, 153C; Crevis: 41BC; crystal light: 99C; Linn Currie: 85C; Dabarti CGI: 139BR; Damix: 79L; decade3d - anatomy online: 105TL; design36: 151B; Designua: 29C, 51BG, 91BL, 91B; Jeanette Dietl: 93BL; DVARG: 71BL; Dzxy: Cover; Ellepigrafica: 49, 61T; Everett Historical: 19T; extender_01: 109C; Ezume Images: 93B; Flukestockr: 131CL&CR; Fusebulb: 113T; Filip Fuxa: 91T; Markus Gann: 103CL; Gen Epic Solutions: 2C, 121C, 149T; Ruslan Grumble: 143L; harmpeti: 123R; Robert Adrian Hillman: 125T(BG); Jari Hindstroem: 81CR; HoleInTheBox: 79TR; Ibreakstock: 61B; Jezper: 49B(BG); Joloei: 79R; Kasezo: 73BR; Sebastian Kaulitzki: 105TL, 109TR, 111TL, 111TR, 151TR; Melissa King: 93BR; Kateryna Kon: 42C, 49C, 56, 98, 109TL, 113C, 151C; Artem Kovalenko: 133C; KonstantinChristian: 123TL; kontur-vid: 147C; koya979: 47C, 93T; KRAHOVNET: 43BG; Le Do: 59T; Lecter: 127C; Leone_V: 29C; Lightspring: 147C; Login: 73L; Lukiyanova Natalia frenta: 85T, 93C; M-vector: 141B&T; Magic mine: 103T; MaluStudio: 17CL&CR; Martan: 151L; Maslenok: 141C; Master3D: 73C; Maxcreatnz: 41C(BG); Maxx-Studio: 143C; Jane McIlroy: 17; Meletios: 61C, 105CL; Mirexon: 143BG; Mix3r:

141C; molekuul_be: 21TR, 21TL, 21BL, 21BR, 57B, 87C, 91CR; Monika7: 17BG, 29B(BG); Monkey Business Images: 63C; Mopic: 37C, 42C, 56, 98, 101C; Darlene Munro: 69B; Naeblys: 51C, 73T; Romanova Natali: 111B; Natykach Nataliia: 149C(BG); Anton Nalivayko: 41C; Nicemonkey: 2BG, 121BG; Nobeastsofierce: 139, 151TL, 151TCL, 151TCR; Ostill: 71TL; Parinya: 73L; Heiti Paves: 119TR; Petarg: 11C(BG), 153C(BG); Phonlamai Photo: 99; Pixel 4 Images: 57T(BG); Pixelparticle: 39(BG); Plan-B: 61T(BG), 139T; Pockygallery: 81BG; Raimundo79: 83T, 127BR, 127BL, 127BCR, 127BCL; Rawpixel.com: 133C; Rost9: 151BC; royaltystockphoto. com: 111TC(BG); sam100: 2C(BG), 121C(BG); science photo: 131T; sciencepics: 87B, 73TC, 151R; CHORNYI SERHII: 133BG; Tatiana Shepeleva: 71BR; David Smart: 141B(BG); Smith1972: 69T; Snapgalleria: 141CT(BG); somersault1824: 37C, 49C(BG), 71TR; Mari Swanepoel: 103BL; Syda Productions: 123BL; T-flex: 131BG; Timquo: 99L; Toeytoey: 109B; Urfin: 73Cr, 143C; VAlex: 139BG; Merkushev Vasiliy: 143L; Vector Tradition: 31C; Vikpit: 127C(BG), 141BG, 149C(BG); Vitstudio: 31C; VLADGRIN: 31C(BG), 71C; Vshivkova: 37TL; Wacomka: 151BC; WhiteDragon: 131C; Wstockstudio: 79BR; Kira_Yan: 17(BG); Yaruna: 29B; Oleksandr Yuhlcheк: 125CT&CB; Oleksandr Zamuruiev: 57TL.

Smithsonian Institution Archives/ Acc. 90-105 - Science Service, Records, 1920s-1970s, Smithsonian Institution Archives: 59C(BG).

U.S. National Library of Medicine: 81TL, 125L&R; Alan Mason Chesney Medical Archives. Victor Almon McKusick Collection: 105BL.

Wellcome Library, London/21BR(BG), 23, 63BG, 105CR.

Wikimedia Commons/ Sandra Beleza et al.: 63B(BG); Belkorin, Wikibob, Quelle: Zeichner: Schorschski / Dr. Jürgen Groth: 149B; Christoph Bock (Max Planck Institute for Informatics): 83C; Dietzel65: 37BG; Filip em: 81TL; Don Hamerman - Institute for Genomic Biology, University of Illinois at Urbana-Champaign: 91; Darryl Leja, National Human Genome Research Institute: 63B; Myriam Létourneau: 63T; Madprime: 25TR, 65TR; Miguel Andrade: 65C; Musée d'histoire naturelle de Lille: 69L; National Human Genome Research Institute: 23CL; Padawane: 69R; Guillaume Paumier: 65CL&BR; RaihaT: 57B; Doc. RNDr. Josef Reischig, CSc.: 123TR; C. Rottensteiner – TiGen: 41T; Dr. Sahay: 147BL; Katja Schulz from Washington, D. C., USA: 119TL; Jawahar Swaminathan and MSD staff at the European Bioinformatics Institute: 147BC; TimVickers: 149C.